FLIGHTS OF DISCOVERY

the earth from above

Georg Gerster

PADDINGTON
PRESS LTD

NEW YORK & LONDON

Library of Congress Cataloging in Publication Data

Gerster, Georg, 1928–
    Flights of discovery.

    Consists of 6 photo-essays from the author's Grand
Design, a translation of Der Mensch auf seiner Erde.
    Bibliography: p.
    1. Aerial photography in geography. I. Title.
G70.4.G47213 910'.1'526 78–6048
ISBN 0 7092 0278 4
ISBN 0 448 22366 X (U.S. and Canada only)

Copyright © 1978 Georg Gerster
All rights reserved
Filmset in England by Vantage Photosetting Co. Ltd.,
North Baddesley, Hants.
Printed and bound in Hong Kong
Color separations by Photolitho AG, Zurich, Switzerland
Designed by Richard Johnson,
assisted by Eileen Batterberry and Sarisberie Designs

IN THE UNITED STATES
PADDINGTON PRESS
Distributed by
GROSSET & DUNLAP

IN THE UNITED KINGDOM
PADDINGTON PRESS

IN CANADA
Distributed by
RANDOM HOUSE OF CANADA LTD.

IN SOUTHERN AFRICA
Distributed by
ERNEST STANTON (PUBLISHERS) (PTY.) LTD.

Contents

Introduction: Beautiful views

*". . . in the plane give him many beautiful views
and don't let the hyenas eat him . . ."*
(from the nightly prayer of my then six-year-old
daughter)

In the last few days of December, 1858, Gaspard Félix Tournachon, alias Nadar, embarked on the first successful photographic flight in a captive balloon; as a precautionary measure he had already applied, a few months before, for a patent on the utilization of aerial photography for the production of maps. On August 23, 1966, at 16.35 GMT, the ground control center in California radioed instructions to the photo-satellite Lunar Orbiter I revolving around the moon to glance back at its home planet. For the first time in history Man saw — even though only through technical servo-eyes — his own earth as a celestial body. After little more than a century, the domination of aerial photography had been superseded by the development of space photography. Prematurely, it seems to me: many possibilities of aerial photography from medium flying heights have yet to be explored.

Certainly, town and country planners, geologists and generals, tax assessors and realtors, cartographers, foresters, agricultural and civil engineers have long been making good use of aerial photography in their respective fields. For them, as for a number of other specialists, it is everyday fare and certainly no cause for emotion. If these specialists and technicians have no wish to get excited, that is their business. As I see it, however, an aerial photograph is a unique vehicle for wonder, vexation, joy, wrath — it never leaves one cold. To those sensitive to visual impressions it is a new training in observation, an unusual school of vision; to the concerned contemporary it is a mirror in which he can see himself as part of his terrestrial environment.

Today, nobody will naively ask a photographer to record the world "as it is." In every photographic effort worth its salt, regardless of whether it is thought of as art, there is hidden an attempt to scratch off the patina of habit that covers everything like a sticky varnish, and to create the world anew. To do this, there is no need to play God: when the ego of the photographer comes to the fore and sets the scene, the result will be just as good as the ego; and that, most of the time, is provincial, artsy-craftsy and pretentious. The aerial view solves this problem at one stroke. The temptation to stage, to manipulate objects, is missing in the first place; moreover, the impossibility of meddling brings freedom: whatever it is that happens to be in the lens will not object. Above all, however, to the observer from above — I am thinking of the almost perpendicular angle, not the church-tower perspective – the world is re-created without tears, as fresh and unsullied as in Genesis.

Philip Gilbert Hamerton, an English art critic of the 1880's defended the customary perspective against the angel's view. He mentally assessed the adequacy of the spatial and aerial picture by imagining the archangel Raphael descending to earth. The only landscapes he accepted in the artistic sense were the ones that could be seen from the ground; somewhat contemptuously, he left the aerial views to the geographers. This opposition can still be met with today, despite the fact that Raphael is now called Gagarin or Armstrong. "The bird's-eye view is nonhuman," a painter friend of mine once reproachfully told me; otherwise nature would have equipped Man with wings. But what about our dream of emulating birds in flight, and the gray matter that enables us to realize this dream? The visual transformation of things that flight bestows on us as a free bonus has always fascinated artists. The Italian Futurists anticipated landscapes of a kind never seen before from *aeropittura*, painting from the airplane — "as if they had just dropped from the sky." Henri Matisse in his old age apparently regretted not having been able to fly when he was young: the view from above in his early years would have saved him the long detour back to his true self. Despite the elevated viewpoint, aerial photography — rather unexpectedly — does not create space, but surfaces: and

6

Matisse wrestled with the organization of surfaces all his life. It is said that the master advised his younger colleagues to fly, for there was no better key to the optical code of the world. A new conception, then: flying not as the shortest route between two crowded airports, but as a shortcut to oneself.

It is hard to get one's fill from above of this second face of nature, the patterns and ornaments that Man, as he lives and labors, coaxes and bullies out of untouched nature. One look at the fields the farmer has laid like a patchwork quilt over valleys and plateaus offers a recipe for becoming an artist without really wanting to. The photographer, however, must be humble. What if he captures a gallery of Poliakoffs – or any other painter – on a single flight! The very fact that he sees them, when he sees them, is the merit of just those artists: they have helped to sensitize us to the forms and patterns that are now revealed in the hitherto hidden beauties of our world.

When my plans for an aerial view of the earth were beginning to crystallize, I still uninhibitedly indulged, as I flew, in a hedonism of the eyes. I was sometimes completely overpowered by beauty – it was always there, lying in wait for me, and I had nothing but an arsenal of cameras with which to confront it. The calligraphy of roads, the graphics of plantations, the unwitting art of salt recovery ponds and the mosaics of small cultivated fields still delight me and tempt me to board planes. But in addition to this beauty "out of the blue," and of equal importance, I am now aware of the information gained from the air. The aerial view by far exceeds the ground-level view in informational content; occasionally it even achieves something like the quadrature of the circle: the volume of information grows with abstraction. Admittedly, first doubts are stirred by the realization that even Man's worst offenses are aesthetically upgraded by sufficient distance. The automobile scrapyard in a natural setting is an eyesore on the ground, but even from kite-flying heights it is transformed into an attractive multicolored design. And as for the profuse, untidy settlement growth that eats into field, forest and meadow: at jet altitudes, if not lower, the eye begins to recognize a gratifying

order in the chaos. Contemplation from a spacecraft redeems the earth from Man completely: to a lunar astronaut it appears as a habitable, though perhaps uninhabited, blue planet. This phenomenon of redemption through distance is the one drawback of an approach that otherwise has only advantages. Distance creates clarity and transforms the single image into a symbol: into an accusation here, a hymn of praise there, a manifesto everywhere. Coincidence turns to fact. On the ground we worry about an inventory of what is, but the lofty contemplation of the aerial photograph shows us also what might be – it is a stocktaking of our chances. Aerial photography x-rays the environments created by Man and reveals the intensity of the ecological give-and-take. It follows Man on his precarious way between foolishness and efficiency, conquest and coercion; manifests Man's conflict between the biblical order to subdue the earth and the necessity, only recently recognized, to submit himself to it. The currently popular condemnation of Man, which sees him as an incurable disease of his own planet, passes judgment without trial. I regard my aerial photographs as the interrogation of the accused; but if they plead at all, it is for one who has built up rather than against one who has destroyed.

So much for the "beautiful views." The hyenas my daughter wanted to deliver me from have accosted in many guises in the course of working on this book: the sloppy plane mechanic, the reckless pilot, the irresponsible aircraft owner – all of whom are exceptions in their trades. I think nothing of boarding a flying machine from the pioneering days of aviation, but I have learned to judge when a vintage plane is a year too old. I refused to be talked into flying in one that was pressed on me for a trip over the Atlas Mountains because of its untrustworthy age and looks. (And with some justification, as it later turned out: the owner had banked on a crash and hoped to collect the insurance money.)

My aerial photographs materialized at the price of a good deal of nervous stress. There was that near collision with a high-speed helicopter over Stonehenge (Plate 6.4), which lies under a NATO corridor used for the testing of new equip-

7

ment. . . . There was that flight above Kyushu in the forefront of a typhoon that had swept the sky clear of clouds. . . . Once we were ready to start at Mosul for the return flight to Baghdad, in the midst of MIG patrols taking off and landing. The control tower suddenly refused us permission to start, a jeep came racing toward us across the runway, and my heart sank at the prospect of the seizure of my films; but the driver smilingly presented me with a bunch of roses – "from the garden of the airbase commander, with his compliments. . . ." During a relaxed conversation on a flight back from the San Andreas Fault, the pilot began to tell me about the heroic days of aerial photography, about carrier pigeons and rockets used as camera carriers when he was a boy – shortly after the turn of the century! When I asked in alarm how old he was, he said he was seventy-four and didn't normally fly anymore, but had to replace his son, who was busy elsewhere. I felt rather queer. I have a lot of respect for old age, but preferably on the ground. . . . On a flight over the crater of Erta Ale (Plate 3.2), dusk surprised us and the nearest airfield – without radio, radar, landing aids or lights – was 30 miles away and 9,800 feet higher, on the plateau. Things looked bad. The pilot, a chaplain and military aviator, had to rely less on his celestial connections than on his flying skill. During the landing approach I read the instruments for him, as he could no longer see them. Finally, in the last glimmer of daylight, we touched down and, in complete darkness, taxied to a standstill on the uneven field. I had made it yet once more. . . . Then there was an emergency landing on the Panamericana, and routine landings on Brazilian airfields where cattle seem to have the right of way. I could go on, but it is not my intention to make the reader's flesh creep.

One's nerves actually suffer less in the air than on the ground, before the first takeoff. For days, sometimes for weeks, one waits for clear weather. Then the pilot of the small plane has to be cajoled into removing a window or a door – or, better still, both. He justifies his reluctance by talking of "safety," but he means "comfort." And almost everywhere one gets involved in battles with officials and authorities for whom aerial photography is synonymous with espio-

nage. The regulations and prohibitions were formulated in the days when there were no satellites observing us uninvited from space. The problems raised by restricted areas cannot always be dealt with in the casual and friendly manner adopted by the authorities in one African state when they warned me against flying into forbidden zones. "As a specialist in aerial photography," the security chief reminded me in a fatherly tone, "you should know best what you are allowed to photograph." When eventually one can take off in a plane that has been stripped to the bone, there is often a security official on board. I have nothing against security officials in general, but a lot against those who object to every shot the photographer wants to take because they are at a loss to understand his passion for innocuous fields, colors and patterns. I have gradually learned, however, how many narrow circles it takes over an approved photographic target to make the watchdog feel sick, and I have discovered that a sea-green security official does not care much about security: he nods apathetically at each and every request. It is particularly satisfying to fly with pilots who see through the same eyes as the photographer, but I have learned to fear those who are too easily infected by my enthusiasm: above the steep escarpment of Bandiagara in Mali we got into an ugly spin because the pilot insisted on taking photographs himself.

Wings, incidentally, are no guarantee that one will not need one's feet. One wintry day ($-20°F$) I took off from Columbus, Ohio, with a pilot who was unfamiliar with the area and we flew around in a vain search for Serpent Mound State Memorial, a prehistoric monument. Finally we landed in a field of stubble and I had to walk half a mile cross-country to reach the nearest farm. At first the farmer was rather nonplussed by the appearance of an alien, dressed in a light suit, who came out of the cold and asked after the Great Serpent. Only after making sure that I had really come from a plane and not from a flying saucer did he offer me coffee and information.

Legwork is also necessary when a place discovered from above subsequently has to be identified. During an exploratory flight along the Big Bend of the Niger, a village of rare beauty

(Plate 1.3) suddenly appeared in my lens; months later I spent many days in the border area of Mali and Niger inquiring about its exact position and name. Labbézanga it is called, and the people clearly remembered the plane that had circled above their heads for hours (or so it had seemed to them). My interest in them and the fact that I had had a plane at my disposal at first aroused their suspicion; they thought I was a government agent sent to arrange for their threatened resettlement. When this misunderstanding had been cleared up, they joyfully hunted for their homes, their friends' houses and the village mosque on the photographs I had taken. Even though these Africans had never seen an aerial shot before, they had no trouble at all reading my photographic plans of their village.

Most of the time I photograph from hired planes, preferably Cessna high-wing monoplanes. Small aircraft guarantee flying in the raw – raw as opposed to overdone, like a steak. For the commercial airlines nothing is too bad (films) and nothing too good (champagne) to make their passengers forget that they are in the air. Anyone who disturbs the routine of film and bottle by flattening his nose against the window hardly gains popularity. But he should not let that worry him. Memorable pictures can sometimes be taken even through the double glazing of high-flying jets. As far as photography from passenger planes is concerned, I remember with melancholy the days before the epidemic of hijacking, when passengers could easily get admission to the flight deck. In the cockpit of the DC-3 there was even a sliding window that could be opened, and now and then an obliging captain would undertake an extra turn for a photographer. I would like to take this opportunity to apologize belatedly to the passengers of two commercial airlines. In the first instance a captain dropped 6,000 feet below the plane's cruising height to allow me to photograph the temple of Abu Simbel in Egypt; in the second, a flight commander kept his plane banking steeply over the rock churches of Lalibela in Ethiopia until my films were used up. In neither case did my fellow passengers know what was happening, and when I returned to my seat from the cockpit very few were able to lift their heads from their paper bags. Had they been up to it, they would gladly have flung me to the hyenas.

Flights of discovery

I. A roof over one's head
(Plates 1.1-1.25)

Aerial photographs showing vegetation on special color film (with colors that are not true to life) enable the biologist to distinguish between sick and healthy plants. This is now common knowledge even in the school classroom. The application of the same principle of diagnosis-from-the-air to settlements is – much to Man's disadvantage – not yet part of the curriculum. The view from above expedites analysis and affords insights difficult to obtain from the ground. For example, a half-hour flight over the Swiss plateau is enough to demonstrate how the cancerous proliferation of today's urban settlements destroys the landscape. In days gone by children with whooping cough were cured by sending them up in planes. Nowadays similar flights should be regularly prescribed as a shock therapy for town planners and citizens alike. "L'avion accuse," Le Corbusier noted in 1935. "The airplane is an indictment. It brings a charge against the city. It also brings a charge against those who control the city." The flier's eagle eye sees the cities as cruel, inhuman creations. It reveals "that men have built cities for men, not in order to afford them pleasure, contentment and happiness, but to make money!" Le Corbusier's vision of "Profitopolis" was directed against the big cities of the nineteenth century, as a contrast to which he cited the new architecture of the machine age.

Forty years on, alas, the airplane still indicts. The accused is no longer just the nineteenth century, but also the architecture and urbanism of the machine age: the worship of function and purpose, the reduction of the need for a home to a mere right to a watertight shelter. . . . The plane accuses. But it can also defend – and acquit. Although Le Corbusier had visited the valley of the M'zab in the Algerian Sahara on foot, it was only after flying over the area that he was overwhelmed by the exemplary nature of Mozabite towns – "a wonderful lesson" he called it. Since then our varied examples of unintentional urbanism, of planning without planners, of anonymous, spontaneous architecture have been increased a thousandfold. We could never have found them, except from the air.

2. Calligraphy of the industrial age
(Plates 2.1-2.23)

The benefits industrial Man derives from the aerial view steadily increase. Airplanes and satellites help in the preparation of plans for engineering projects – dams, roads, pipelines, railroads, etc. The plotting of maps and charts would be unthinkable today without aerial collaboration, as would the stocktaking of field and forest, area planning, the search for raw materials, the preservation of the environment, the gathering of up-to-date information on the state of our planet, the movements of sand in the deserts and of ice in polar regions, the variations of the snow line, the sediment transportation of large streams, the migration of cloud banks, and so on. Remote sensing not only uses the window that is open to us in the visible part of the electromagnetic spectrum, it also "sees" in the infrared and radar ranges. These Argus eyes in the sky, first used for military purposes, now carry out their espionage for Man. The information they provide is the preliminary for global management. This should not be forgotten.

The accompanying pictures are a study of industrially active Man himself and of his appendix, the man of leisure, the consumer of spare time. The examples selected are calligraphies of our technical civilization, but they are more than just aesthetically interesting. Some undoubtedly belong to the controversial images of a changing era, and have triggered some widely differing reactions. The road through primeval forest, for instance, is certainly a manifesto – but of what? Only a short while ago the heart beat faster at the thought of pioneers advancing through the jungle: now it reacts with shock. Or the concrete lianas of the Pacific metropolis of Los Angeles – a great engineering achievement, but to what end? The most traffic-oriented of all cities was for decades a monument to the internal combustion engine, symbol of a great conquest, until suddenly its pollution became notorious and even bumper stickers began to celebrate the days when sex was dirty but the air was clean.

3. Flight over the Afar
(Plates 3.1-3.11)

Landwards the Afar triangle is bounded by the precipitous escarpment of the Ethiopian plateau and the terraced slope of the Somali tableland. Seawards it is flanked by the Red Sea and the Gulf of Aden. This desert, three times the size of Switzerland, belongs mostly to Ethiopia but overlaps in its southeastern corner with the territory of French Somaliland and the Somali Republic. The few thousand nomads traversing it – members of the Danakil tribes – can hardly be considered a population. The catastrophic droughts of the last few years have reduced their numbers drastically and driven them to the edge of their territory, where the climate is a little more favorable. *Afar*, meaning "the free," is the name the Danakil give themselves and has recently led to a new designation for their triangle of land.

For the traveler with no concern for the geosciences the Afar desert is an inferno, particularly in the north where it sinks below sea level. This is the planet's thermal pole: in the summer the mercury column creeps towards 140°F in the shade. Bleaching skeletons mark the way through this desert without hope or horizon; salt grinds underfoot and salty sweat stings the eyes. Then come oceans of dunes, lava slag and cinder fields, geysers that periodically spew steam and water, fumaroles and solfataras, and air reeking of sulfur. A vision of hell – and gaudy as sin: in Dalol yellow salt flowers grow out of a green brine, the chlorates of potassium and iron blossom into unearthly gardens. All in all, it is the apex of inhospitality.

But this is not the only reason that scientific research has been held up for so long in this part of the world. Danakil custom requires that young males should decorate their brides with the genitals of slain enemies. The fear of castration deterred even those who might have been willing to risk their lives, and geologists and geographers preferred to avoid the Afar desert. For the last eight years, however, this African valley of death has attracted swarms of now unquailing scientists, for the Afar triangle might prove to be the Rosetta Stone of a new geographical discipline, plate tectonics. (As it happens, palaeoanthropologists have also recently begun to see their Mecca in the Afar desert. The discovery of "Lucy," a unique prehistoric skeleton – 3 million years old and 40 percent intact – makes the Afar a "cradle of Man.")

The science of plate tectonics rehabilitates Alfred Wegener and his hypothesis of the continental drift. Yet, according to this new Disney-like conception of the constant changes taking place on the face of the earth, the continents no longer plow through the earth's crust but travel as stowaways on huge plates constantly forming and disintegrating. Plate tectonics has identified six major plates of planetary proportions and, depending on the theories of individual scientists, hundreds of minor ones. Processes taking place in the earth's interior keep these plates in constant motion. Through the V-shaped median valleys at the summits of submarine ridges magma oozes forth, forming new oceanic crusts and pushing the plates apart. Where such plates collide (in the zone of the deep sea trenches), the heavier plate is forced down into the molten rock of the earth's crust.

An example of these tectonic uplift processes is to be found in the Afar desert, where just such an oceanic mountain, whose magma springs add new material to the plates, has gone aground. Here it is possible to study dryshod phenomena that otherwise can only be investigated by diving and drilling from research vessels. Iceland is a similar lucky strike for geoscientists. The Afar triangle, however, is of particular interest to the scientist because several seam zones of the earth's crust meet here at a single point.

The frequency of earthquakes and intense volcanic activity are evidence of the tectonic lability of this area. Beneath the desert there is enough heat to supply the whole of Africa with electricity, and the red glimmer of the volcano Erta Ale lights up the night sky of the northern part of the desert. The "Smoking Mountain" (as the Danakil call it) and the Nyiragongo (Kirunga) in Zaire are at present the only volcanoes on earth with a permanent lava lake. Erta Ale is the only instance of a continuously erupting volcano in the median valley of a dried-out oceanic ridge.

An aerial photograph I took of Erta Ale in 1965 (the first known aerial view of this volcano) later helped volcanologists to assess the extent of its movements: in the space of a few years the lava in the two craters rose 460 feet and eventually overflowed.

In the Afar desert three fault systems meet in the form of a star: the trough of the Red Sea, the continuation of the Carlsberg Ridge in the Gulf of Aden, and the East African rift valley. A large number of French, Italian and German geologists as well as some British, American and Ethiopian scientists are interested in this "rift star." According to the Belgian volcanologist Haroun Tazieff, spokesman of a French-Italian group, the trough of the Red Sea and the continuation of the Carlsberg Ridge in the Gulf of Aden are two sections of the same rift valley, which bends round in the Afar. His theory is based on arguments derived from surface geology. His colleagues, who plead for two independent systems, substantiate their disagreement with geophysical measurements, mainly seismological and obtained from explosions. It is certain that the Danakil depression in the north of the Afar triangle is an arm of the Red Sea which has been cut off by uplift processes. Here the salt lies to a depth of thousands of yards, and the potash deposits in this evaporating pan are estimated at millions of tons. On the lava floor of this enclosed sea basin – dry now for tens of thousands of years – corals can still be picked. Many peculiarities of this triangular rift valley system are yet to be explained. Wegener had already noted that the southwestern corner of Arabia and the African coast lying opposite fit each other like two pieces of a jigsaw puzzle if the Afar triangle is disregarded, but even today it is difficult to explain it away on scientific grounds. Parts of the desert lie well above sea level; and some horsts of continental crust are troublesome stumbling blocks. Just as the bursting apart of the earth's crust between Africa and Arabia once created the Red Sea, in a few million years the Afar triangle will be flooded again (Tazieff calls it the "Erythraic Sea" in anticipation). But opinions on the present stage of its "oceanization" differed greatly in the spring of 1974 at an international Afar Symposium in Bad

Bergzabern in the Palatinate. Tazieff and his

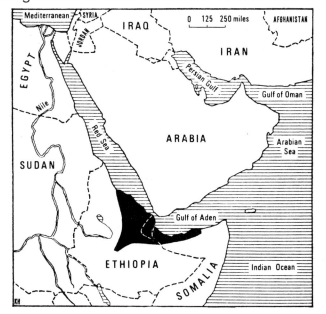

group already seemed to see the magma rising in all the clefts and fissures, preparing the way with its basalt for a future ocean. The German geophysicists, however, were more reserved; their measurements suggest a thin but still intact continental crust.

There are still many such unsolved riddles. From a flight across the Afar depression in 1971 I brought back a photograph of a swirling labyrinth of salt blocks. Later on, again from the air, I tracked down two further examples of such labyrinths, albeit badly disfigured by erosion. Geologists who had worked for years in the vicinity of this salt lake had, to my astonishment and their own, never come across these mysterious formations. For the moment they disagree as to their meaning – in fact, they are completely at a loss.

4. Flight over Nebraska
(Plates 4.1-4.13)

The Midwest, granary of America (and other parts of the world), appears to the traveler on the ground as a monotonous expanse broken only by church and water towers, grain elevators and silos. Dull? Certainly not from the air. The view from above transforms this landscape into one of the most visually exciting in the USA. And makes it clear that farmland is nature shaped by the hand of Man, that the farmer can – by purely technical production methods designed to maintain or increase his yield – change the graphic face of the landscape again and again in the course of thirty or forty years.

The optical delights of Nebraska derive from two sources. The first (and older) consists in the measures taken by American farmers, warned by bitter experience, to prevent the erosion and blowing away of the topsoil. The second (more recent) has resulted from the development of irrigation techniques enabling farmers to make better use of the fertility of the soil.

The most important of the conservational techniques is called strip cropping. On terraces, along the contour lines of cultivated areas, or at right angles to the direction of the prevailing wind, strip cropping prevents the loss of humus, stores humidity and increases harvests. In areas of low precipitation, where dry farming is practiced, the zebra landscape has emerged naturally – partly because fields have to be left fallow in summer to restore the moisture balance, partly because of America's right-angled surveying: the individual sections could only be divided into strips. In other parts of the country federal and state authorities have successfully encouraged strip cropping since the devastating dust storms of the 1930's. This practical concern for the soil has thus introduced visual tensions and beauties into the former monotony.

The farmer, then, as artist or designer? Does he appreciate the changing garb of his fields? He certainly has the means of admiring his creations from above, for many farmers use their own planes to go to work in the fields or shopping in town. The fact that here and there real strip orgies are celebrated makes one think twice about it. One can check all the factors that affect the original decision – for instance, direction of the wind, availability of mechanical equipment, steepness of slopes, run of contour lines, re-

13

quirements of a particular crop, etc. – but the calculation never works out exactly. There is always a remainder, and that might well be creative irrationality.

The strips of the forties and fifties have now been joined by the irrigation plants of the sixties and seventies. They are characteristic patches on the face of America's new agricultural landscape. There can be no doubt about the success of these circular irrigation roundabouts in the arid regions of the U.S.A. Every flight from coast to coast reveals new irrigation works in the states of the Midwest and, beyond the Rocky Mountains, in California and the states of the Pacific Northwest. This irrigation system – *center-pivot irrigation* in the terminology of the specialists – was invented more than twenty years ago. With Patent No. 2604359 the American Patent Office protected Frank Zybach's "Self-propelled Sprinkling Irrigation Apparatus" on July 22, 1952. The inventor and his first licensees had to wait a long time for the success of their product, but for a good five years now the number of those who consider July 22, 1952, a day of revolution has been growing rapidly. At the present time some 12,000 plants are operating in the U.S.A. and two dozen manufacturers are looking for more clients.

The jet perspective reduces these irrigation turntables to the size of toys. In reality they are of considerable proportions, each an automatic farm in itself. The standard model can irrigate some 160 acres and larger models irrigate quite a bit more. The biggest – near Yuma, Colorado – covers a circular area of 520 acres.

The swivel arm with its spray nozzles rotating about the center consists of a number of sections. Each separate section is individually driven. If it moves too fast or too slow, a control signal switches the motor on or off and brings it back into place. In the majority of installations the discharge pipe of the swivel arm is supplied with groundwater from a borehole near the center. Some plants also utilize water from rivers and lakes. The sections of the rotating arm have hinged joints so that it adapts to uneven terrain and is not stopped even by terraces. The technically most advanced systems run on tractor tires (causing minimal damage to the soil) and are electrically driven. The farmer selects the desired speed from the control desk at the center of the roundabout, transmission being infinitely variable from 20 to 200 hours per revolution. He also sets the water supply at the well necessary for the desired density of irrigation, and the rest is up to the machine. It can even be programed to sprinkle only a particular sector of the circle and then switch itself off or turn around – for instance when the farmer is growing several crops with different moisture requirements in one irrigation circle, or when the circular irrigation area includes a large obstacle.

The advantages of these irrigation plants are obvious: they are fully automatic and require only one-tenth the labor of conventional irrigation systems. One man is capable of operating up to thirty center-pivot irrigation systems. Irrigation specialists also stress the economical water consumption.

As could be expected, the costs of purchase and installation of these systems are high. In Nebraska an apparatus of standard size, together with a well 100 to 200 feet deep, sets a farmer back approximately $30,000 to $35,000. In the Midwest stupendous bankruptcies have occasionally occurred when farmers have failed to adapt their irrigation roundabouts to the conditions of the soil – a danger inherent in all irrigating. Such inexperience resulted in catastrophic compression of the soil. Today the lesson has been learned, and irrigation systems that were originally reserved exclusively for valuable crops can today be economically viable even when the irrigated area is only pasture. Most banks now lend money on these plants without mortgage securities.

One farmer in the northeastern corner of Nebraska operates no fewer than a hundred irrigation systems. There are good reasons for the concentration of the plants in this state – Nebraska in fact owns one in three of the American total. The fact that the inventor himself lives there hardly counts for much. The real reason is the abundance of groundwater in Nebraska and the liberality inspired by this wealth. At present, drilling for water in Nebraska requires registration only and not an official permit.

5. Biblical cities and sites from the air
(Plates 5.1-5.12)

Palestine was among the first places to be photographed from the sky. During World War I airborne photographers of the German-Turkish Detachment for the Protection of Monuments operated over the Sinai region and the Negev. Since then, the airplane – or sometimes just a balloon or kite – has become indispensable for archaeological work, particularly in a country where high places have always had a religious significance. Like archaeologists today, the kings of that time were aware of the advantages offered by an elevated standpoint. But how could we today, without aviation, look down on the buildings they left behind – at the acropolis of one of Solomon's chariot cities, or Herod's castle and tomb?

The flight into the past is more rewarding over Palestine than anywhere else. It opens up an incomparable source of historical knowledge, which is illuminated in turn by the Bible. A flight over Palestine is memorable, too, for the beauties of the landscape. The hills of Samaria shine like pure silver. Galilee appears as a carpet of flowers. In the distance the snow-covered Hermon towers over Mount Tabor: "Tabor and Hermon shall rejoice in thy name" (Psalm 89). And the mountains of Judaea sing the praises of the farmer's toil.

6. Monumental question marks
(Plates 6.1-6.14)

How far from the earth can an astronaut travel without losing sight of all signs of earthly life? And how near must a visiting astronaut come to be able to recognize signs of life on our planet?

The earth's signals, like those of a radio star, penetrate a few hundred light-years into space, and could be received by extra-terrestrial intelligences even if they are no further advanced in communications technology than modern earth dwellers. For the last few decades at least, the irregularities of the earth's radio signals must have revealed to possible neighbors in the Milky Way that this is an inhabited planet.

And the reflected light? To the astronauts on their way to the moon, still in the terrestrial field of gravity, the earth appeared like a marble on the black velvet of space – colorful but cold, with clouds, oceans and land masses suggesting the possibility of life but supplying no evidence whatever. Carl Sagan, head of Cornell University's Laboratory for Planetary Studies and spokesman of the American exobiologists, has studied thousands of photographs obtained by meteorological satellites for indications of life. He summed up his investigation in a challenging paper that asked the question: Is there life on earth? As long as two points lying only 100 meters apart cannot be perceived separately from space, the traces left by Man's activities cannot be satisfactorily distinguished from geological features. Only when resolution is better than 100 yards can such traces be recognized for what they are. But even then there may be misunderstandings about the type of life assumed to exist on our planet. Sagan once said, only partly in jest, that Martians might well take the automobile to be the principal form of life on earth. After all, the environment is changed to meet its requirements; it moves, eats and ejects the products of its metabolism; it is looked after by an army of two-legged slaves, among whom it regularly selects victims for ritual slaughter. ...We have to take a step nearer to the earth, and then suddenly the patterns of living creatures are joined by their monumental question marks.

The Plates

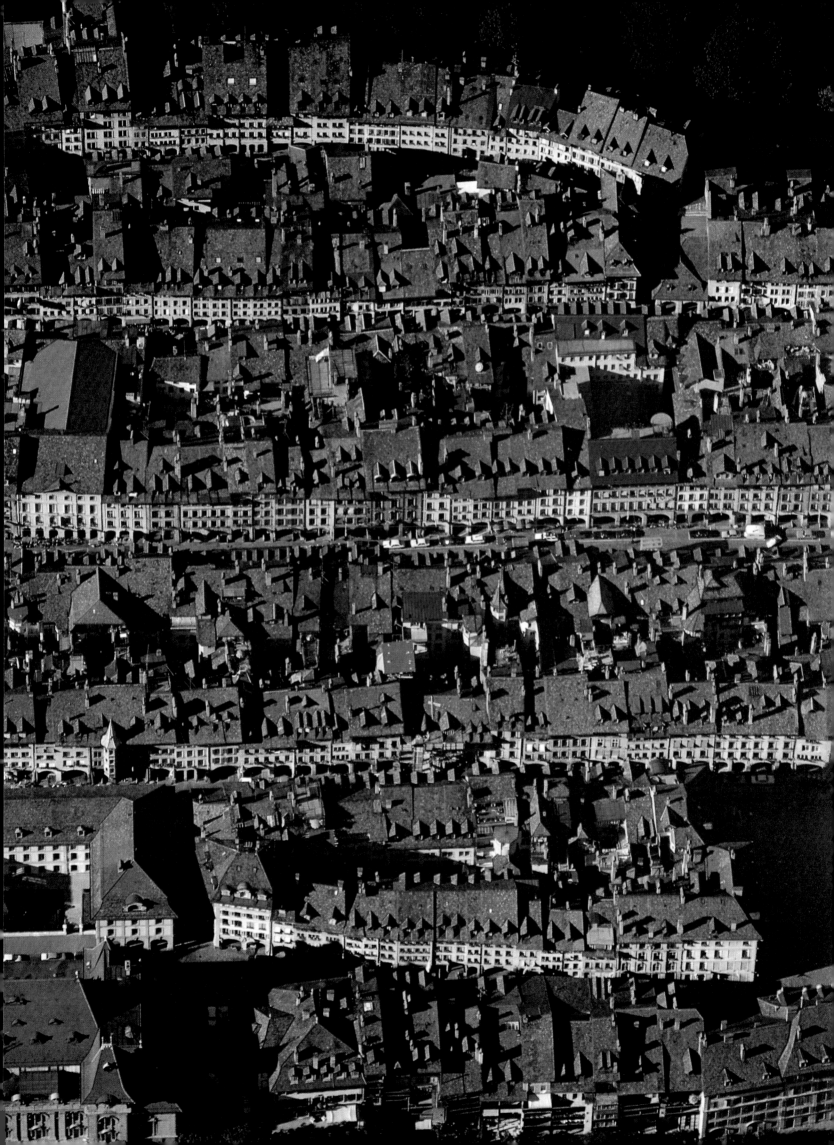

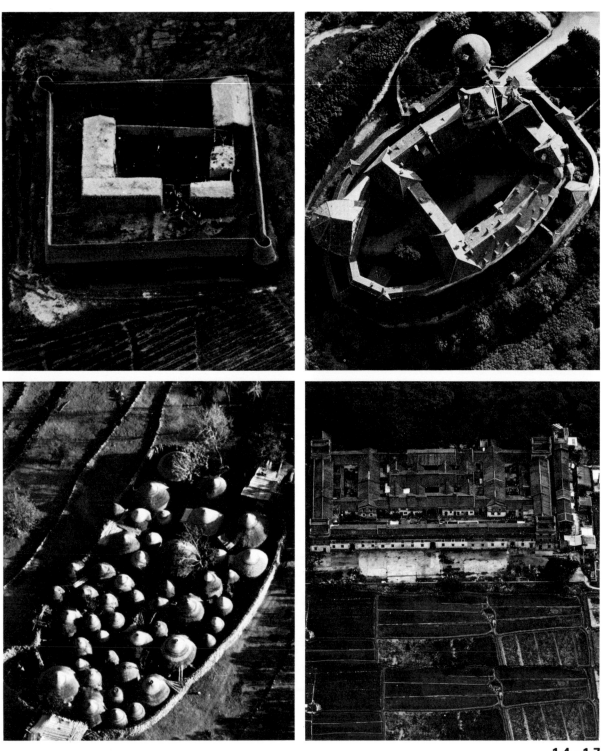

1.11

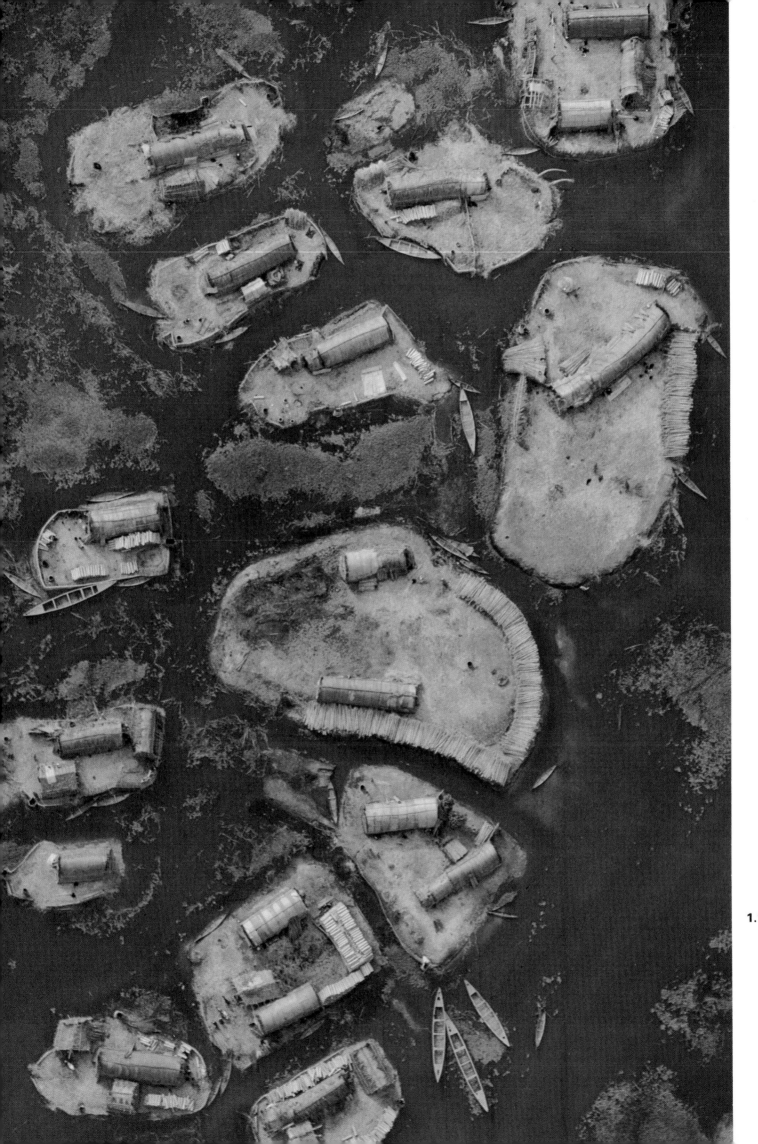

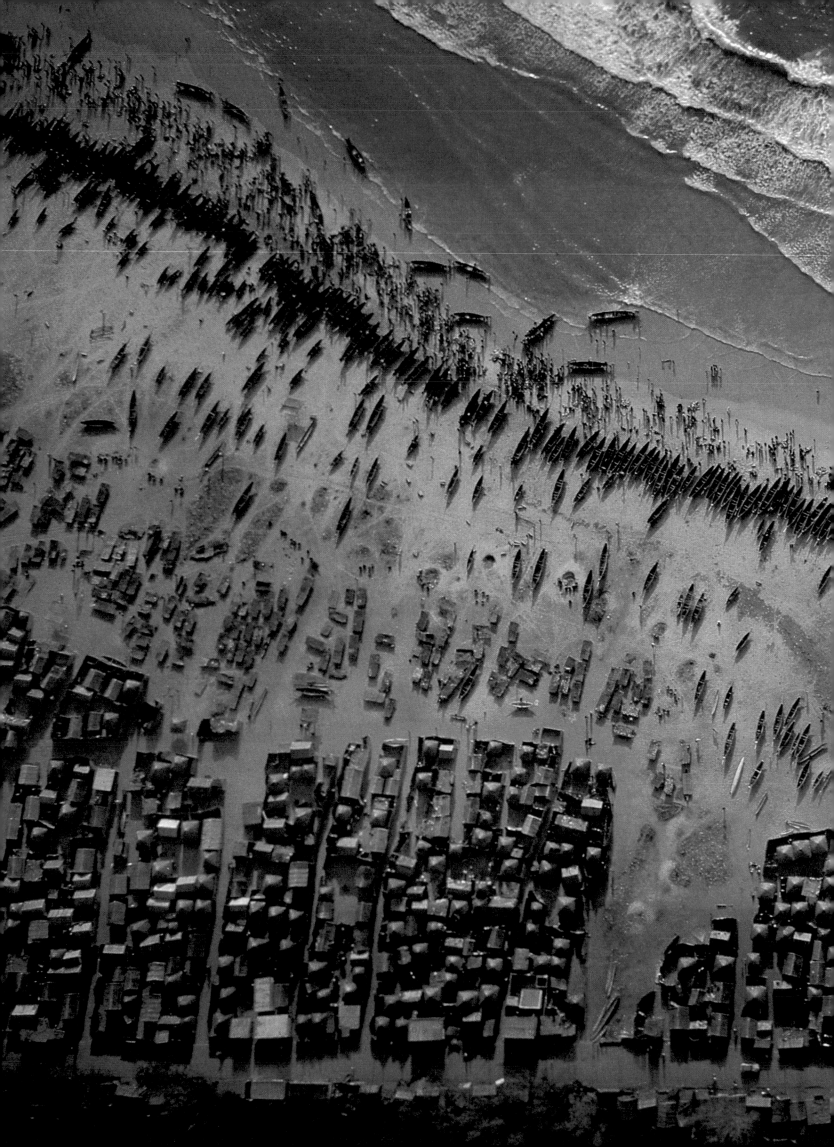

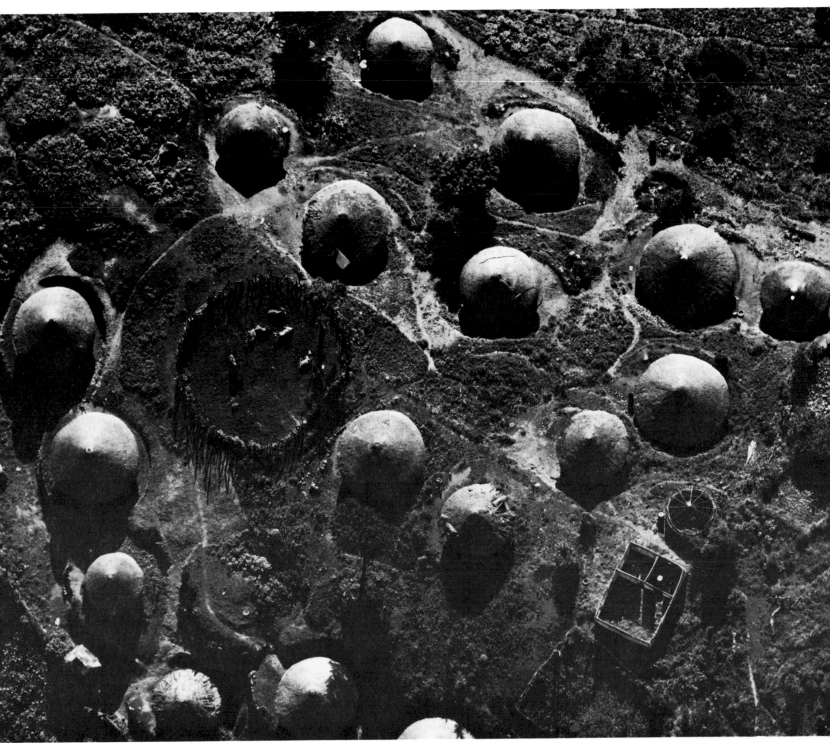

1.25

2.2

2.3

2.4

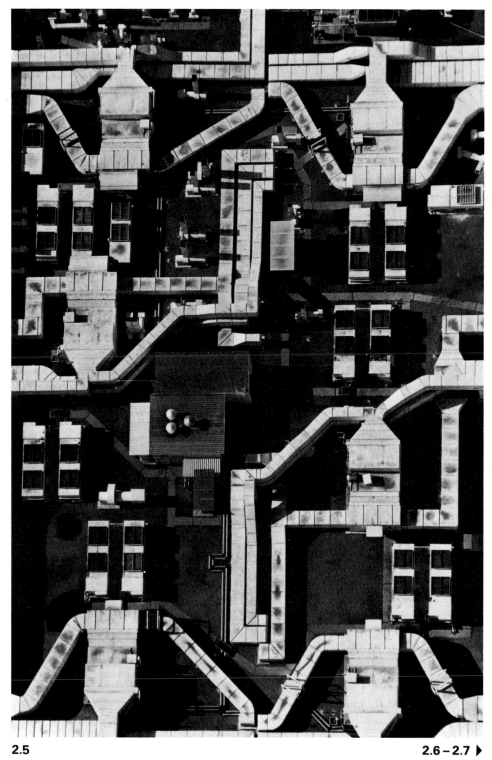

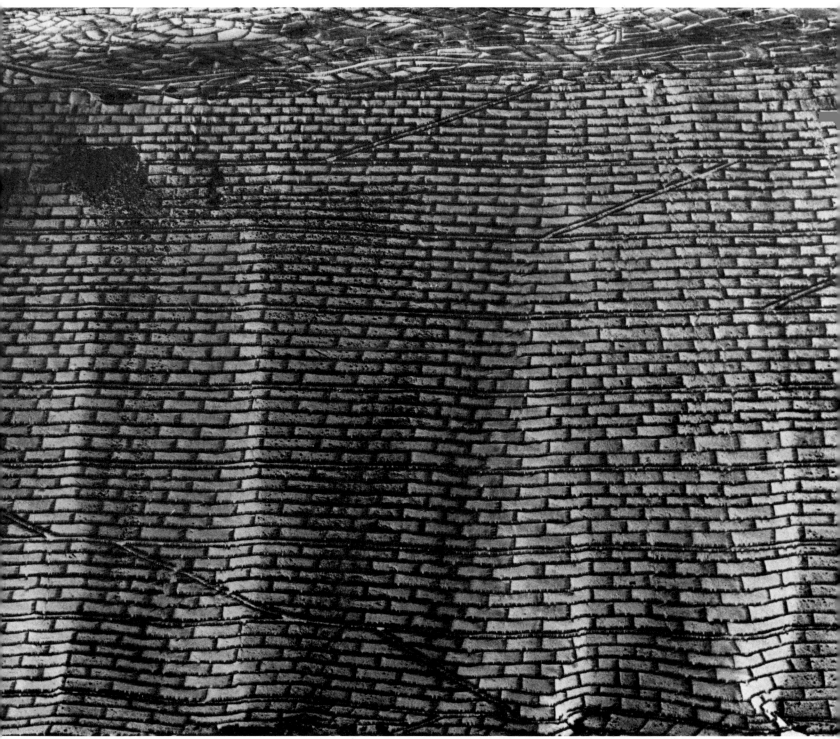

2.10

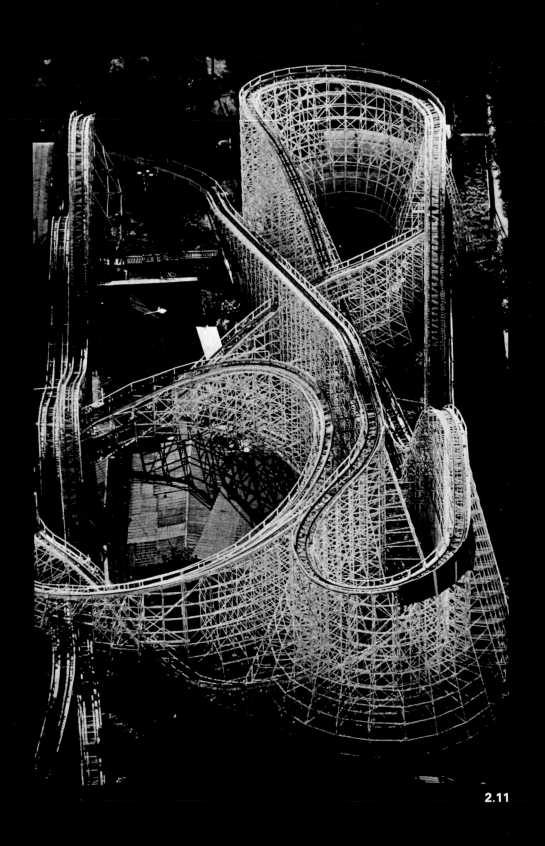

2.11   2.12

2.13 – 2.14 ▶

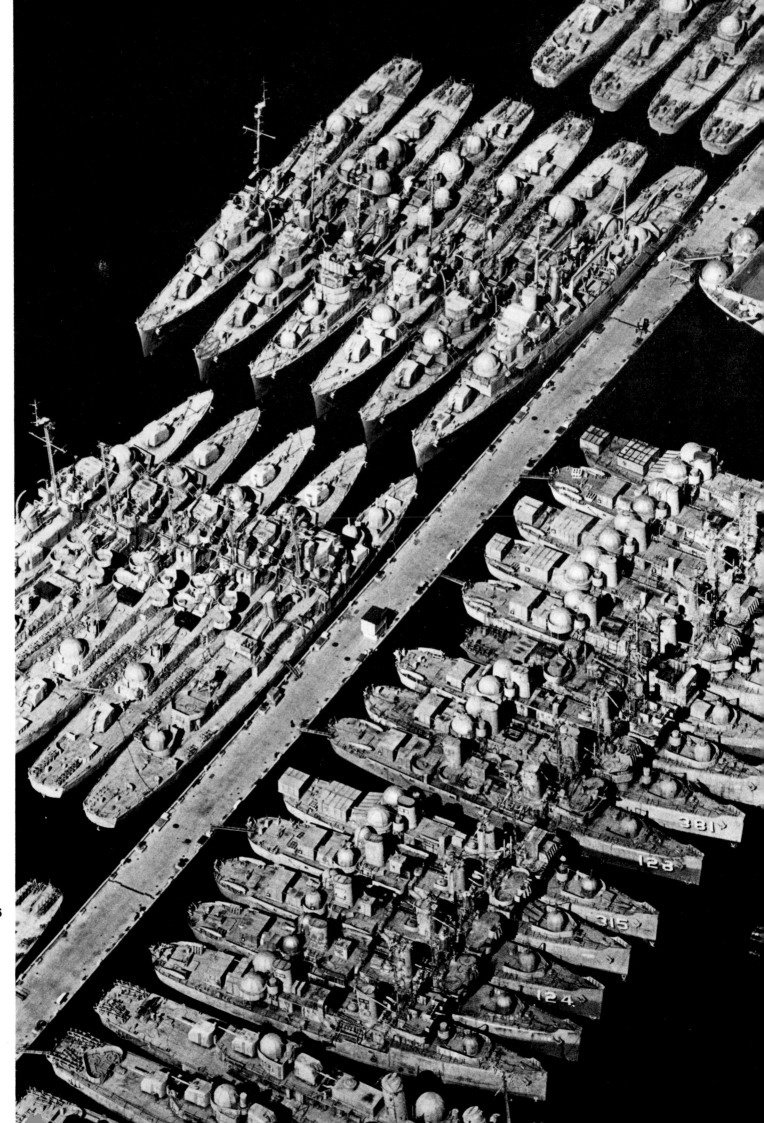

2.17

2.18

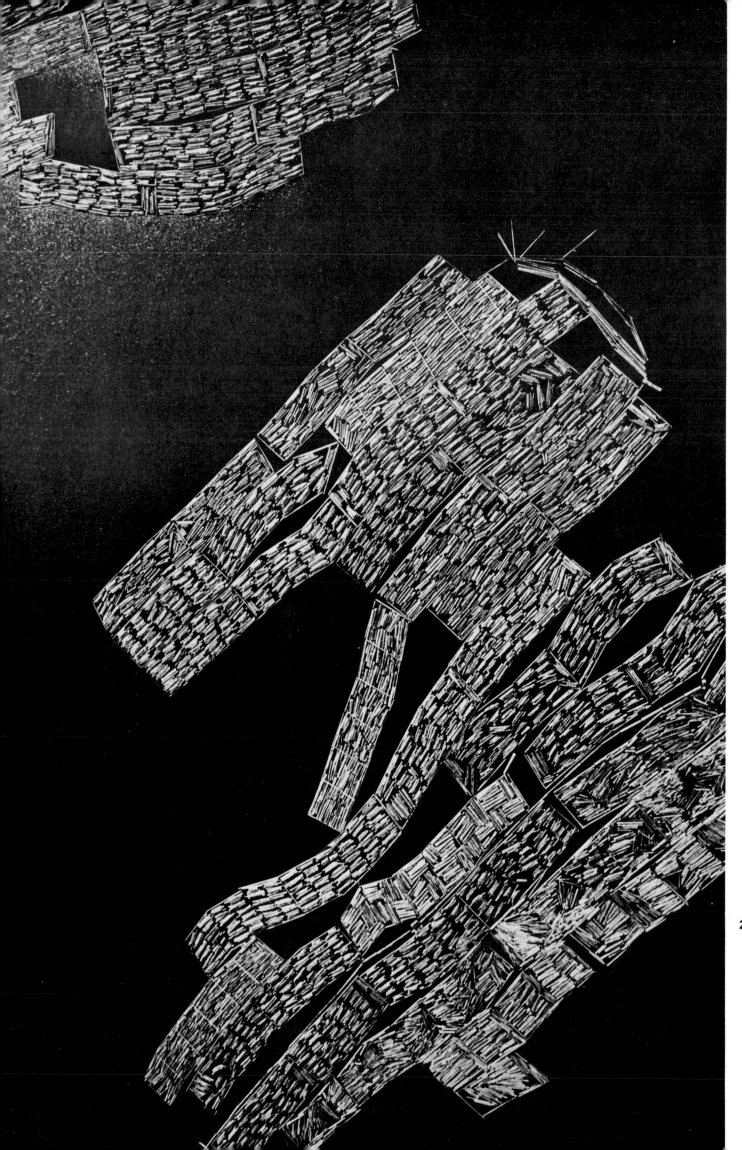

2.20

2.21   2.22

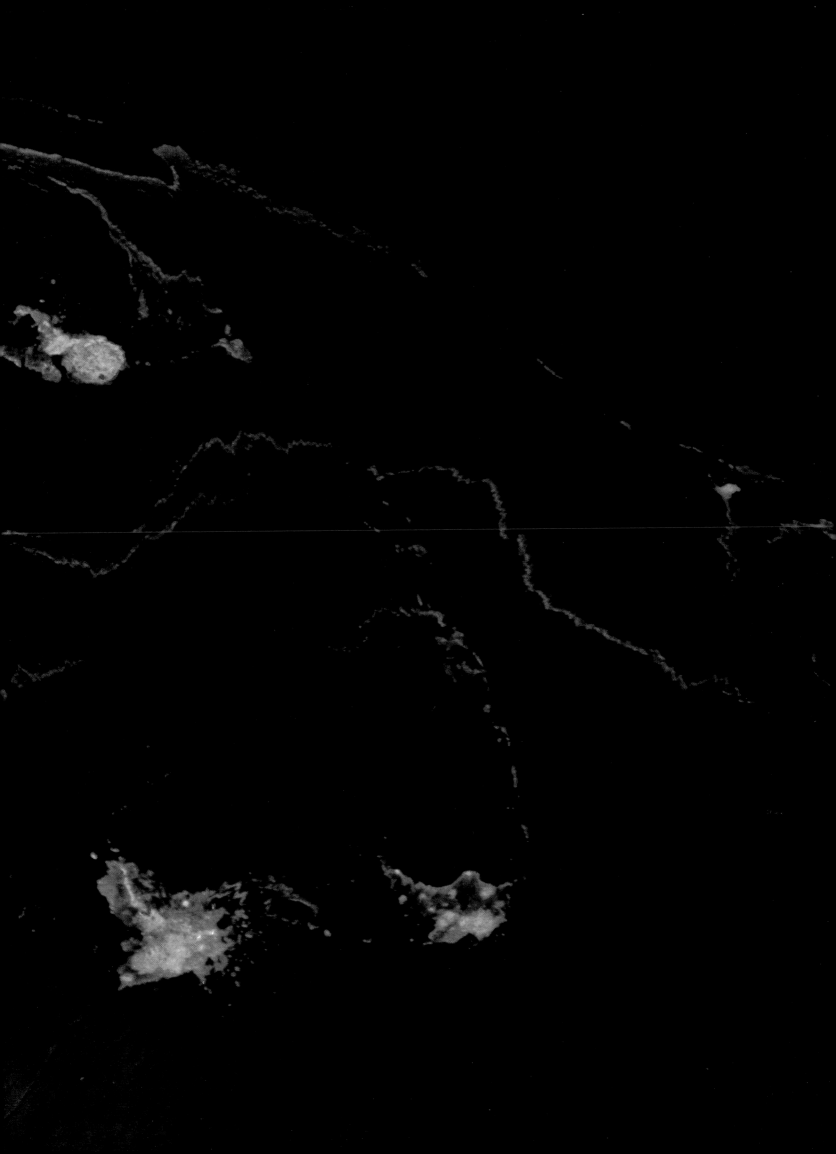

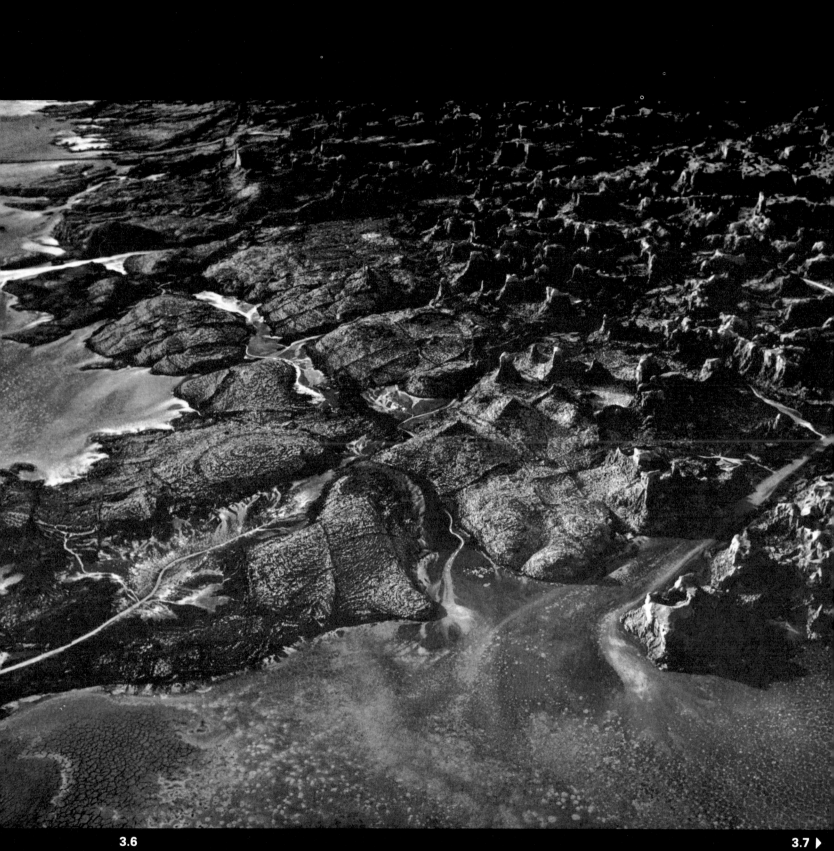

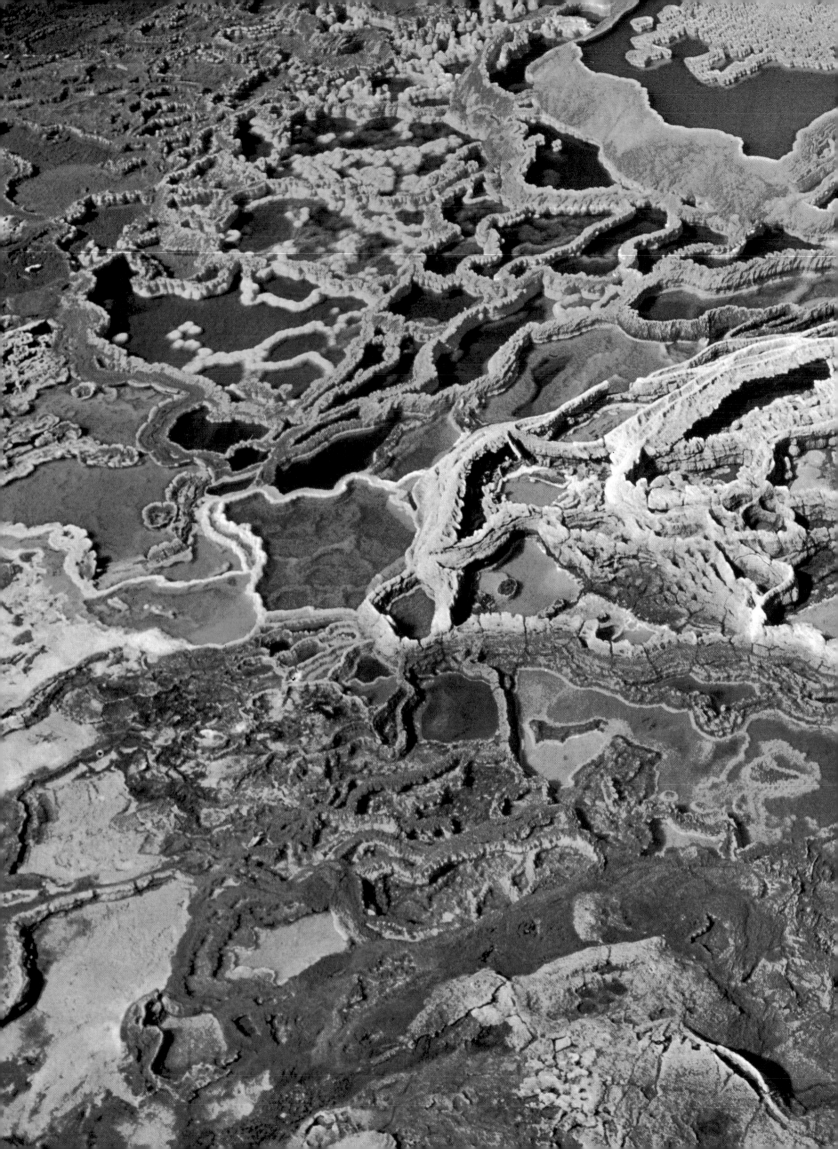

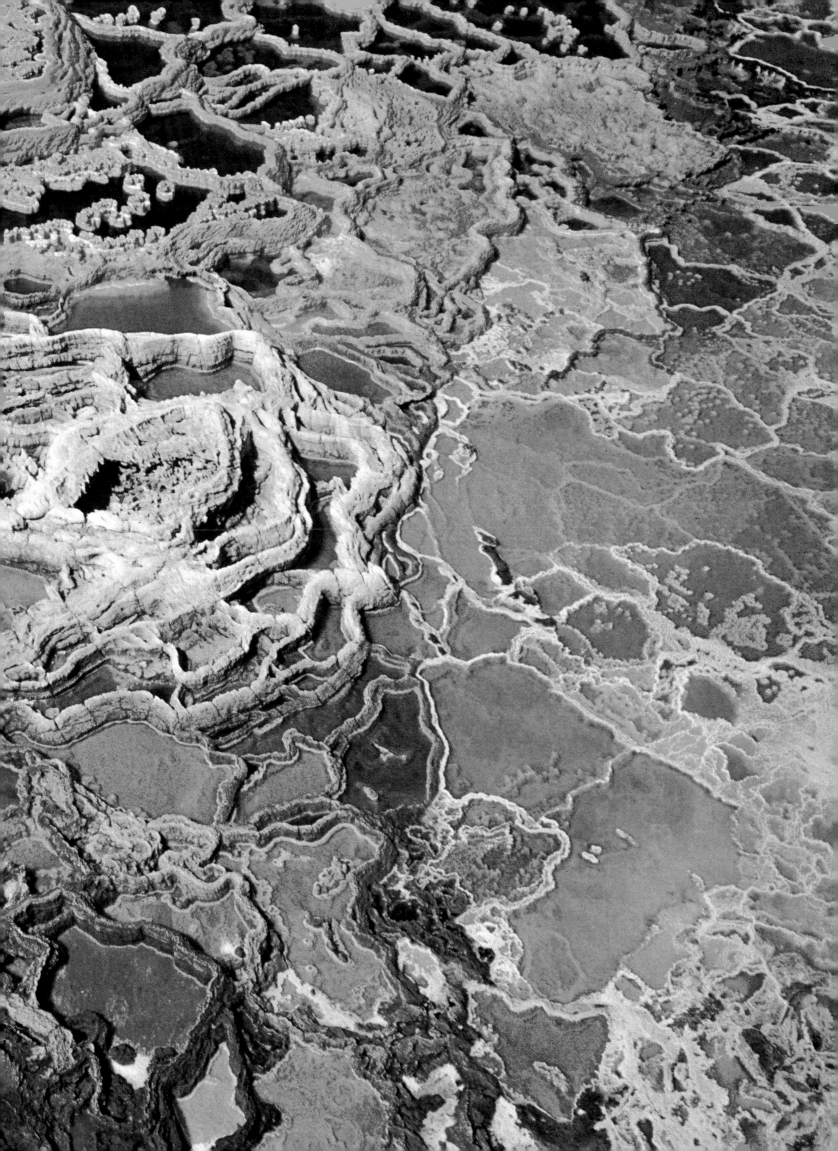

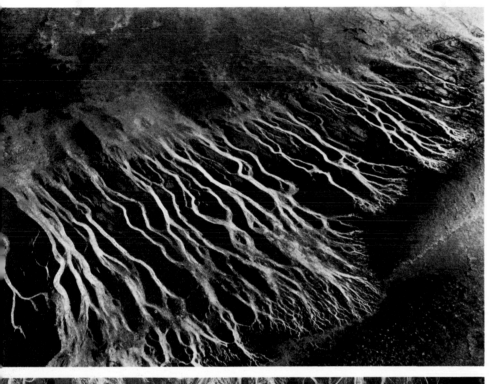

3.8 – 3.10

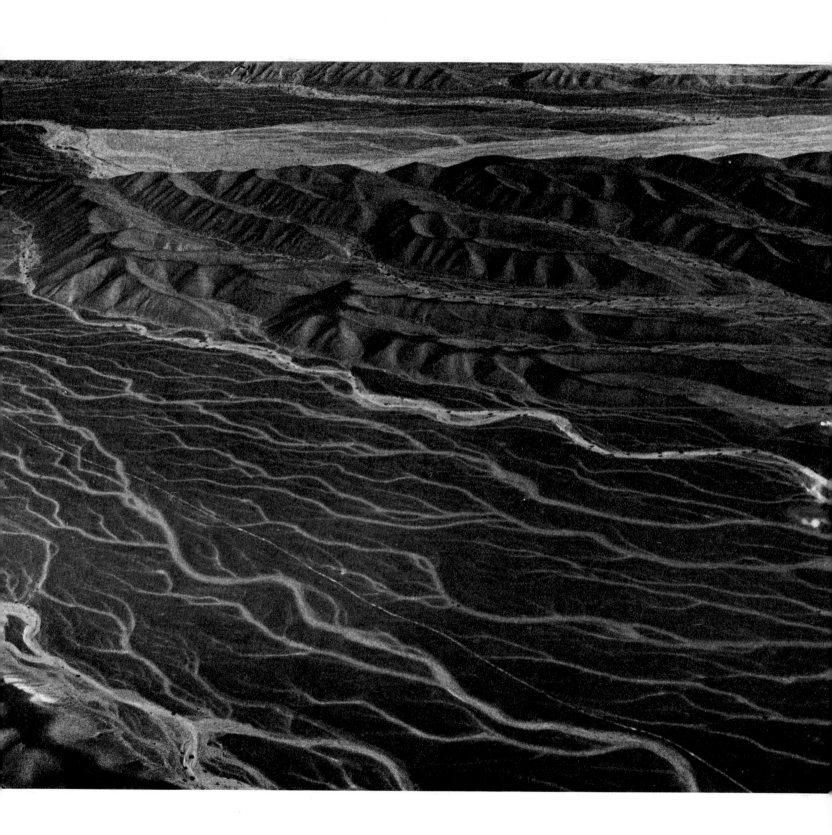

4.1

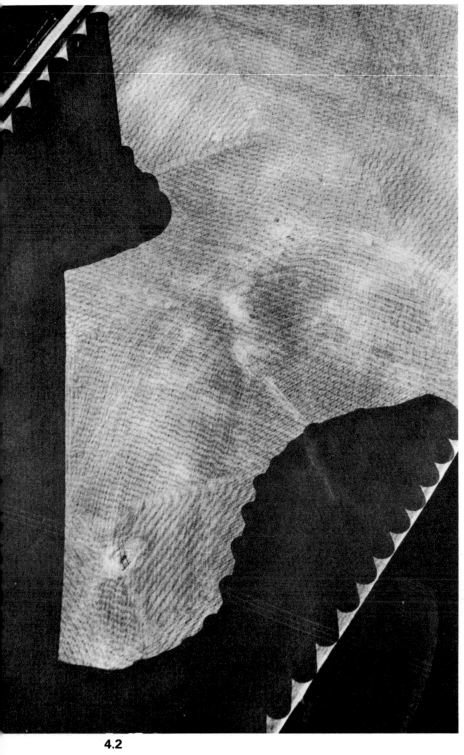

4.2

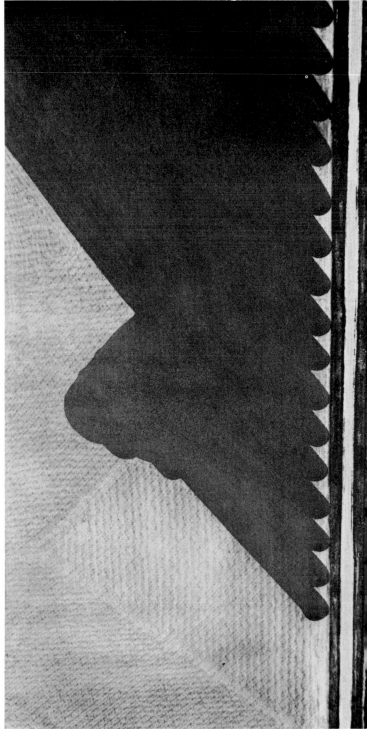

4.3

4.4

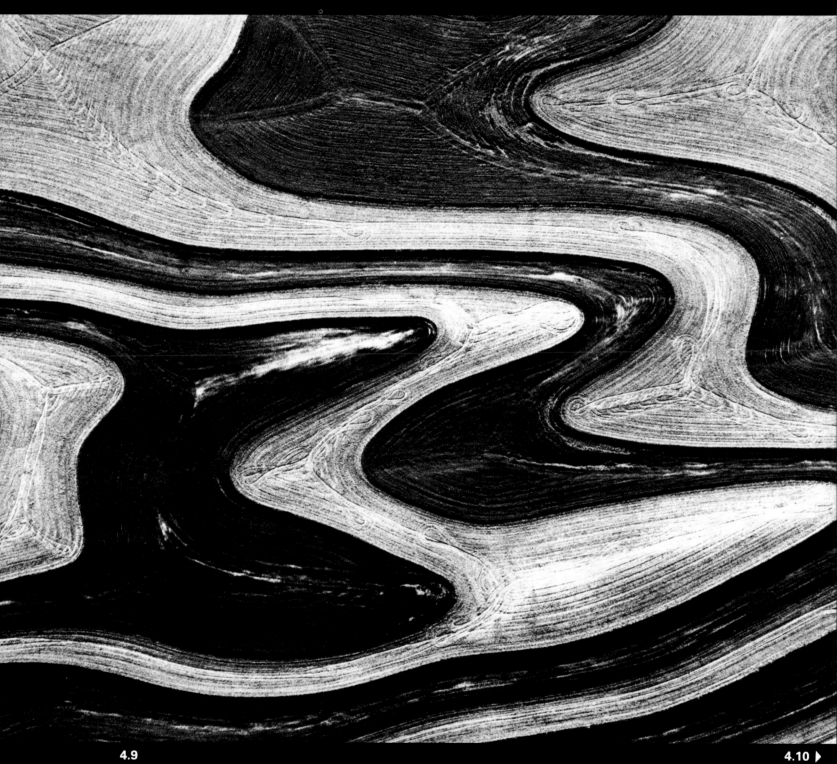

4.11

4.12

4.13

5.1

5.2

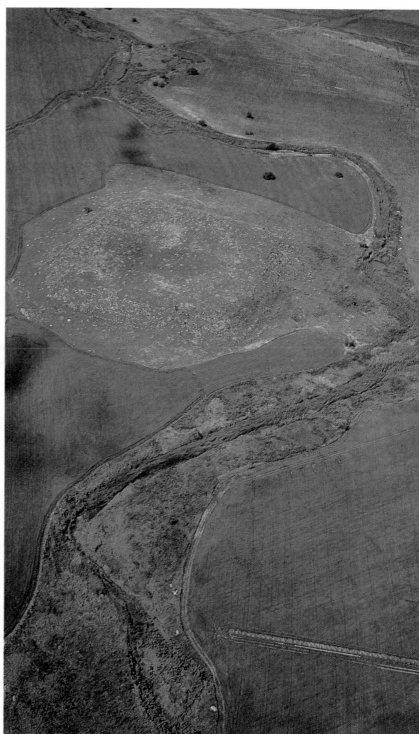

5.3

5.4

5.5 ▶

5.6

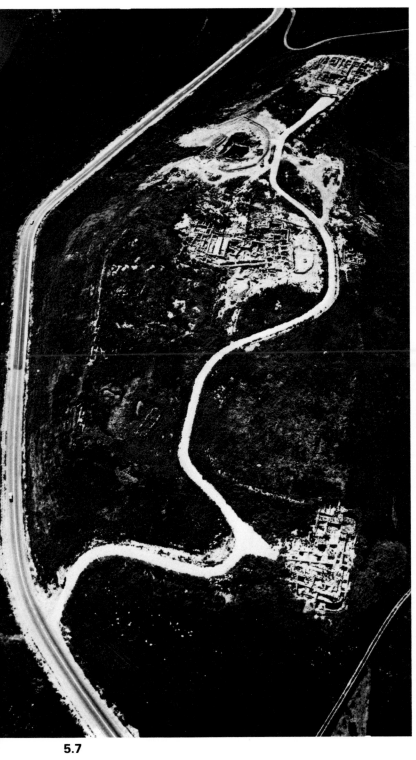

5.7

5.8

5.9 ▶

5.10

6.1

6.2

6.3

6.4 ▶

6.5   6.6

6.7

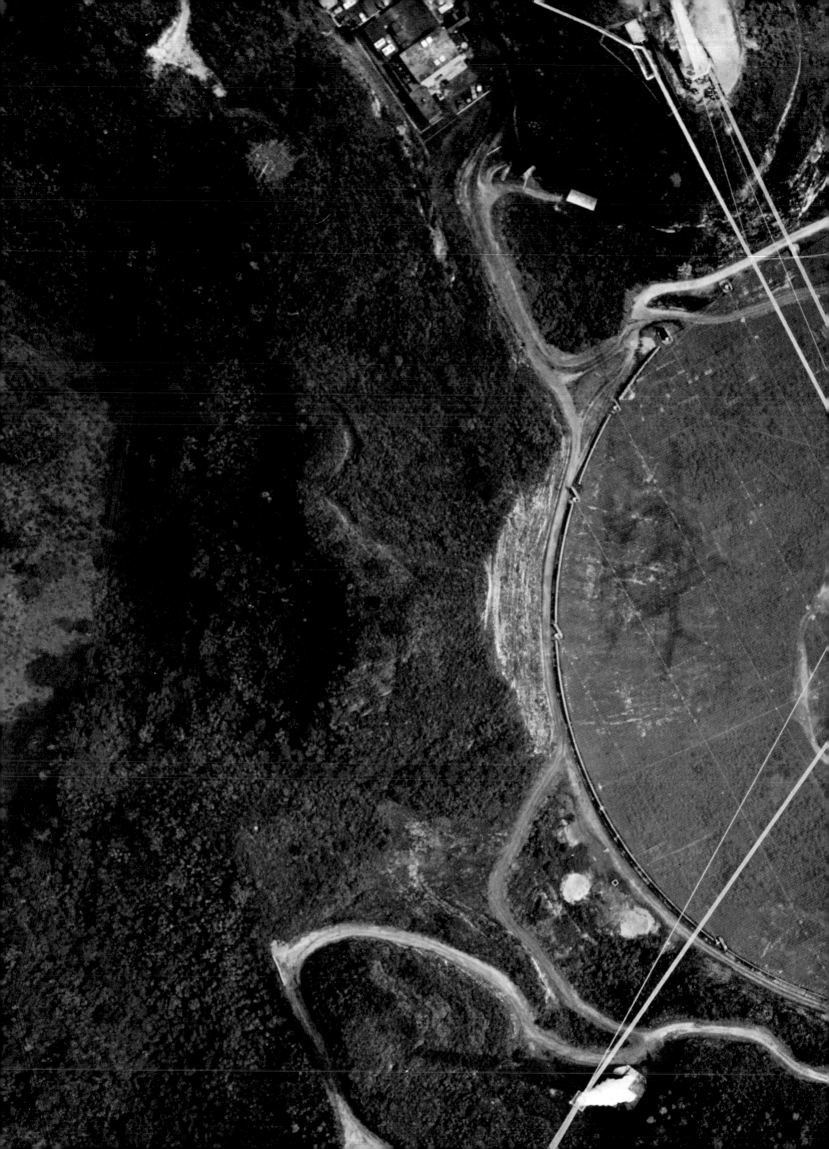

6.11

.12 – 6.13 ▶

Caption notes on the plates

1.1 A masterpiece of medieval urban architecture: the old part of the town of Berne, Switzerland. It represents in unequaled purity and maturity the urban pattern of an axially structured town with a street market. In Berne the wider main street that served as a marketplace – Kram – and Gerechtigkeitsgasse, one of the most astonishing and congenial street systems anywhere – is flanked by smaller streets running parallel to it. The eaves of the houses are aligned and their gables form long rows. In front of the ground floors of the houses are the famous arcades: citizens were permitted to build these arcades before their houses and to extend the house façades over them; the ground under the arcades, however, remained public property. Berne is one of a whole family of towns founded by the dynasty of the Zähringers on both sides of the Rhine. The urbanist Paul Hofer, the leading authority on Berne, praises it as the "most compact and consistent creation of town planners, in the twelfth century at any rate."

1.2 The oasis town of El Oued, capital of the Souf in the Algerian Sahara. A showpiece of "architecture without an architect" – the dream of every town planner. . . . Vaults and domes covering all living and service rooms shed the wind-blown sand, which settles as a sound-damping layer on the streets. The house with courtyard is the unit of the town, which grows on the building-block system without any waste of land. El Oued, which has 50,000 inhabitants, is the most important market and administrative center of the Souf. A hidden stream flowing under the sand determines the form that is taken by date palm cultivation in this area. But the architectural originality of the townscape is also connected with this stream. Water rising by capillarity evaporates before it reaches the surface, and the salts dissolved in it are thereby precipitated. Thus in one zone are formed banks of almost pure gypsum, which, when roasted, can be used as mortar and for making bricks in molds. In another zone the gypsum becomes encrusted with quartz sand to form sand roses and later *lus*, the local name for a hard, durable brick which sets extremely well with the gypsum mortar. This explains the brick construction of the Souf houses, which is very unusual for the Sahara. The clay generally used for building is not available in the Souf, and timber is if anything even rarer than elsewhere in the desert – hence the domes on the houses, the simplest form of roofing where timber is too costly. The workers erect these semi-cylinders and hemispheres by hand, disdaining any frames or shuttering. Although in this part of the world the dome is the privilege of religious buildings and of palaces, in the Souf the poorest and least holy of men lives in this respect like a marabout or pasha.

Only the very well-to-do can afford a two-story house with a common or garden flat roof!

The village of Labbézanga on an island in the Niger, Mali. 1.3 The storehouses for millet and rice wind through the village landscape like strings of beads. These amphora-shaped storage bins, some of them as high as the huts, are filled and emptied through an opening at the top; stone slabs and fragments protruding from the body of the storehouses make them easier to climb. The villagers, who belong to the Songhai tribe, still live for the most part in the traditional round adobe huts with hemispherical straw roofs. (The other traditional house form of the Songhai is shown in Plate 1.21.) Even in Labbézanga, however, the typical square house of Islamic-Arabic architecture is gaining in popularity: to live in one boosts the owner's social standing. A border village between Mali and the Republic of Niger, Labbézanga should have been moved to the river bank several years ago – a measure dictated by the government in Bamako for no more legitimate reason than official convenience. The villagers, looking at the ruins of a fortress of the once powerful Songhai empire on the island and remembering the defiant spirit of their ancestors, refused to obey the order from Bamako, and the government did not act upon its threat to evict them by force. Allah be praised for both! I know no more beautiful village in the whole of black Africa.

A farm equipped for self-defense to the south of Baghdad, 1.4 Iraq. These small fortresses, which are occasionally intended as places of refuge for whole settlements and mostly belong to the sheikh, are known locally as *qalaa*. The rectangular walls have bastions at two corners so as to open the flank of attackers to those defending the farm.

Burg Kreuzenstein, a fortress near Korneuburg, Austria. A 1.5 Burg Kreuzenstein already existed in the twelfth century but was destroyed by the Swedes during the Thirty Years' War. About the end of the last century it was rebuilt as a model of a Romanesque-Gothic fortress. Medieval building material gathered from the whole of Europe was used for the purpose.

A walled farmstead of the Kirdi near Rumsiki, in the 1.6 Mandara Mountains, Cameroon. Huts and barns are huddled together for protection inside a stone wall. Such farmstead forts – known as *sarés* – lie like hedgehogs in a landscape of rural toil and travail, synonyms of the fear that drove the Kirdi up into the comparatively safe mountains. As animists (Kirdi means "unbelievers"), they owe both their fear and their name to their Mussulman neighbors who wrested the fertile plains from them.

1.7 A Chinese farmstead in the hinterland of the Kowloon peninsula, Crown Colony of Hong Kong. In these large farmsteads whole clans live together in very confined conditions. The ground plan of the farm is classical; comparable architecture can be found in pictures dating back to the Han dynasty (206 BC to 220 AD). Today the defenses serve only to ward off demons and evil spirits.

1.8 A village on the lake of Kainji, a storage lake of the Niger in Nigeria. The villagers are busy pounding millet and cooking breakfast as a new day begins. The picture hints at the delicate balance between the "public" and "private" domain, one of the basic features of African living. Stake and mat fences offer no protection against armed attack; theoretically rather than materially, they keep out "nature," safeguarding the clan's territory against wild beasts and the evil eye of strangers. A loud call for "authenticity" is now heard in Africa. No return to genuine African traditions can be complete without the acceptance of native forms of dwelling and settlement; in spite of the not very durable building materials, they greatly facilitate a deeper understanding of human communities.

1.9 Saint-Malo in Brittany, France. The Bretons, a saying has it, are born with sea-water around their hearts. If that is true anywhere, it is in Saint-Malo. The Malouins are famed as explorers and seafarers: it was one of them, Jacques Cartier, who first reached the St. Lawrence and gave the territories he took possession of for France the name of Canada. And it was the Malouins who, as corsairs – officially sanctioned freebooters with letters of marque from the king – harried the English, Spanish and Dutch on the high seas. This territory smaller than the Tuileries produced many great men; its most famous citizen, Chateaubriand, seemed more satisfied than astonished at this fact. The Malouins are a proud and self-assured race, and their island town is an architectural gem which they have worked on since 1144, when it became an episcopal seat. Recent reconstruction – true to history both in spirit and appearance – has cleared away the extensive destruction wrought by World War II in this hotly contested town.

1.10 A suburb of Seattle, Washington – a paradigm of American suburbia. The well-tended residential estate, a monument to the good neighborhood (everybody in the same income bracket), and a breeding-ground for status-symbol culture and the seeds of racism. Among all the arguments advanced for the suburb, the automobile is the most dangerous: four-wheeled mobility intensifies the centrifugal forces that are bringing the hearts of American cities depopulation and death.

1.11 Downtown Los Angeles. The car-conscious Californian megalopolis has become the embodiment of the anti-town: a swarm of parking lots in search of a city. In the background the section of the Santa Monica Freeway that carries the highest traffic density (175,000 cars per day on eight lanes). More urban freeways will be found in Plate 2.3.

Islands and barrel-type houses of the Madan in the swamp 1.12 wilderness on the lower reaches of the Euphrates and Tigris, Iraq. Where the two rivers meet lies the waste of reeds and water known as Hor, for centuries a place of refuge for the flotsam and jetsam of the tribes, with many pre-Arabian usages and customs. The only means of transport of the swamp-dwellers are the *meshhufs*, beaked boats with a high waterline, which they punt with great skill through the reed thickets. They live on their buffalo herds and on the sale of rush mats. Their semicylindrical dwellings are made wholly of reeds, and the islands themselves are also built on a foundation of reeds and other canes. In the course of the years, fed with the dung of the water buffaloes and the waste from the huts, these islands grow till they reach a height of several feet above the winter low-water level. Hydraulic engineering projects on the upper courses of the rivers, particularly the evening-out of high and low water in the interests of irrigation, today represent a threat to the thousand-year-old way of life of the Madan.

Dwellings on piles in Lake Nokoué, Benin. Lake Nokoué – 1.13 also known as Lake Ganvié, after the biggest village – is the lagoon behind the sand bar on which Cotonou, the country's capital, stands. Fifty thousand people live as fishermen on the lake. They use net and harpoon, but also practice a kind of fish farming by putting out twigs for the fish to spawn in. Fishing is the preserve of the men, who sell their catch to their wives, who in turn sell the fish (mostly smoked) on the market. The lake dwellings are hardly older than the present century. After a channel had been cut through the sand bar in 1895, the freshwater lagoon became salty, and the increasing salinity made irrigation of the land impossible. The farmers were forced to become fishermen and moved out into the lake.

Kayar, Senegal. For seven months of each year life here is 1.14 centered upon boats and fish; for the remainder of the year the village at the water's edge is deserted. In December it grows into the biggest fishing settlement on the West African Atlantic coast, only to die out again in June, at the end of the fishing season. Beyond the barrier of foaming breakers, rich catches attract the fishermen. The art of riding the waves in pirogues seems almost to be inborn in these West African fishermen. Yet the woman and children still await the return of their menfolk with anxiety when, coming home each evening, fathers, husbands and sons negotiate the dangerous wall of surf.

114

1.15-1.16 Zulu settlement in the "Valley of the Thousand Hills," near Durban, South Africa; and Nilotic farmsteads along the Sobat, a tributary of the Blue Nile, Sudan. The round hut with a conical roof is the typical farmer's house in black Africa. The herdsmen usually lives in a hive-like or hemispherical hut without cylindrical walls. Originally the Nilotic tribes were solely cattle breeders, and the Zulus were strongly influenced by them. But the form of dwelling chosen by both of these tribes reflects the growing importance of agriculture in their lives.

1.17 The Edmundshof farm on Parndorf heath, Austria. The manor belongs to the Cistercian abbey of Heiligenkreuz. Parndorf heath, a sheet of gravel in Austria's Burgenland, is wind-swept and dry. On the occasion of a renovation seventy years ago, the farm was surrounded by a double circle of trees as a windbreak and was replanned with dwellings, stables and outhouses so far apart that a fire would not burn down the whole farm.

1.18 Oil palm plantation in North Sumatra, Indonesia. The settlement of the plantation workers forms a bright square. It is the germen of a company town and belongs to the prescribed settlement forms in which self-interest and goodwill – here that of the plantation company – are intermingled. In the foreground and middle distance young plantations are visible. For technical harvesting reasons the palms, which begin to bear in their third year, are only allowed to grow to a height of fifty feet, which takes twenty-five years. One twenty-fifth of the plantation area, therefore, has to be renewed each year. In North Sumatra, at the present time about 180,000 hectares (nearly 450,000 acres) of land is under oil palms, with 143 trees per hectare. One palm yields around 225 pounds of fruits per year. Every harvest worker is expected to cut a ton of them every day in his block – about fifty bunches of fruit, each with hundreds of plum-sized drupes rich in oil.

1.19 Parts of the Konso settlement of Buso in the province of Gamugofa, Ethiopia. The Konso are among the most interesting of vestigial tribal societies in southern Ethiopia. They live in three dozen permanent settlements, the compactness of which are a feast for the eye of the town planner but a nightmare to epidemiologists. The population density of Konso settlements, which are towns rather than villages, is attained nowhere else in Ethiopia (and probably nowhere in the rest of Africa). The modular unit is the single farm with huts for living, sleeping and cooking, a storehouse and a stable. The single farms form quarters on the basis of family and clan, community organizations and ritual groups. Each quarter has at least one *mora*: a festival ground with a men's house in which married and single males sleep so as to save energy for war and hunting through sexual abstinence.

Part of the city of Nagasaki in Kyushu, Japan. In the period 1.20 during which Japan cut herself off from the outside world (1639 to 1859) Nagasaki played an important role as the only port in Japan where Dutch and Chinese vessels were allowed to call. Thus, it sometimes acted as an exchange center for Western goods and Western science. One of the most nefarious products of the latter, baptized "Fat Boy," was later responsible for the fact that today Nagasaki is all new: on August 9, 1945, at two minutes past eleven, an atomic flash obliterated 150,000 people and their city. Today heavy industry is concentrated in Nagasaki, which exports ships (see Plate 2.20). Since World War II a weakness for roofs with colored tiles, particularly blue, has spread over the whole of Japan.

A Songhai village near Timbuktu, Mali. The more than 1.21 500,000 Songhai live on the Great Bend of the Niger. They fish, cultivate a narrow strip of land along its banks, and harvest wild rice. Where mobility is required by their way of life or the terrain, they use exclusively a hut made of matting which resembles the shell of a tortoise. In this they follow the shoals of fish or escape high water by moving to elevated land. But even when they settle down permanently they are loath to renounce the habits of their seminomadic past. Those who can afford to keep up with progress may acquire an adobe house with a flat roof, or at least a rectangular walled courtyard. Within it, however, the tortoise-shell huts still appear, either as accommodation for guests or for the hot summer nights. The building of these matting houses is women's work, and they are therefore the wife's property. A man who leaves his wife also loses his home. Plate 1.3 shows another type of dwelling used by the Songhai.

A camping town near Copenhagen, Denmark: the city 1.22 dweller's frustrated dream of unspoiled nature.

Mobile homes in Long Beach, California. At bottom left, 1.23 part of the concrete canal grandiloquently labeled "Los Angeles River." The trickle in it only lives up to its name in flash floods. For the privilege of parking his motorized dwelling the owner pays a monthly rent. Most mobile home estates, however, are not satisfied to be mere parking lots for portable houses: they embellish themselves with communal facilities such as clubhouses, saunas and swimming pools, recreational centers, golf links and gymnasiums. This extra glamor glues their tenants to the location, once chosen. One estate may offer personal wine cellars, another will hold a Rolls Royce at its clients' disposal for transfers to and from the local airport. Their calculations seem to be working out. Every year Americans buy over half a million mobile homes, and are proud to uphold the covered-wagon tradition. But once they have found a good site, they stay there – in fact, they move house less frequently than apartment dwellers.

1.24 Junks in the port of Aberdeen, Hong Kong. A hundred thousand people live on 21,000 houseboats and lighters stationed in the natural harbors and typhoon shelters of the Crown Colony. They are born on their boat and they die on it – very rarely do they set foot on *terra firma*. Aberdeen has the lion's share of Hong Kong's floating population. For the great feasts of the Chinese calendar almost 13,000 junks will meet in its port, and whole clans are anchored in an intimate circle, bow to bow. The inhabitants of the junks make their living by fishing, by small transport services within the harbor area or along the coast, and as traveling tradesmen. Because of its high, flat stern and unfavorable ballast distribution the junk is not very well equipped for the open sea. Junk enthusiasts will not admit this, but even the Hong Kong Tourist Office – certainly no enemy of things picturesque – warns visitors of the disadvantages of these boats out of official concern for their safety.

1.25 Brasilia – the reverse of the medal. One million people have been attracted by the nucleus of the new city, but such multitudes had never been reckoned with. A ring of squatter settlements (here known as *invasões* – "invasions") grew up around the city, which had been conceived as an architectural and urbanist showpiece. Glass, steel and concrete on the one side; iron sheeting, roofing felt, old crates and rags on the other. The battle order in which the "invasion" advances makes mockery of town planning by aping some of its principles. Despair packed in corrugated iron? Oscar Niemeyer, Brasilia's star architect, resigned in the face of these hovels, shacks and sheds: "In the end, Brasilia was a town like the others, a town of rich and poor, unjust and discriminating." Contemporary urban sociologists, however, make distinctions. They agree that not all "invasions" (and not all *bidonvilles*, shanty towns, *barriadas* and *favelas*) are slums. Slums are dead ends of misery, traps of desperation – including the comparatively well-heeled ghettos of North America and the noble slums of Europe. By contrast, squatter settlements, despite their unattractive building materials, may also be places of hope, scenes of a counter-culture, with an encouraging potential for change and a strong upward impetus.

2.1 A road through the jungle in the Brazilian state of Mato Grosso. President Juscelino Kubitschek, founder of Brasilia, complained about the lack of pioneering spirit in his compatriots who, he said, clung like crabs to the overpopulated shores of their huge country. He called up the spirit of the *bandeirantes* – those forest rangers, treasure hunters and adventurers who, in the seventeenth and eighteenth centuries, set out from São Paulo and penetrated the unexplored interior. "We must march west, turn our backs on the sea and stop staring at the ocean as if we were permanently thinking of sailing away." Kubitschek dreamed of Brasilia as the focal point of a network of roads which were to lure the Brazilians away from the coast into the enormous wilderness beyond. The tempo and spirit with which reality is today catching up on and overtaking Kubitschek's vision are alarming, to put it mildly. A gigantic road system 9,300 miles long for the development and colonization of the Amazon basin and the Mato Grosso is either already built, is in the course of construction or has at least been fully planned. Hardly is the bulldozer epic of the Transamazonica from Recife to Rio Branco completed when interest turns to the 2,500-mile *Perimetral Norte* from the Atlantic to Colombia, which runs through the jungle north of the great river. The last primeval expanse on earth, the planet's richest reservoir of oxygen and fresh water, trembles under tires and caterpillar tracks. But the sin against ecology is not the whole story: the steamroller of civilization brutally crushes the forest-dwelling Indios. They are massacred by the white man's diseases; his graders and planers, sometimes mistaken for super-tapirs, drive them from their homes ever deeper into the jungle. His Colt and his knife permit no waste of time. Colonization degenerates into colonialism of the worst kind. Brazilian Indianists have been courageously condemning this secret genocide in the virgin forest for the past fifteen years. There are even utilitarian arguments for the preservation of these forest-dwellers. They have survived in this green hell of poisonous and stinging plants, of snakes, bloodsucking ticks and insects, and they have learned to live healthily in it – an achievement made possible by their own medical, botanical and pharmacological knowledge. Though otherwise backward, they could therefore very well make a cultural and even economically valuable contribution to the taming of the jungle. But this consideration of expediency has so far been ignored in the conquest of the Amazon region. The frankly conquistadorian attitude of the modern *bandeirantes* often turns the dream road into a nightmare.

2.2 Ways for human beings: the steps of the Piazza di Spagna in Rome. A masterpiece of early eighteenth-century architecture, they connect the Piazza with the Trinità dei Monti church on the Pincio. The flower vendors at the foot of the stairway are part of the traditional setting of square and steps. The flower children at the tops of the flights hawking their homemade wares – belts and buckles, rings and bangles – are a new accent.

2.3 Ways for automobiles: city freeways in Los Angeles. A model for the control of traffic in agglomerations? Or writing on the wall, warning us against choking our cities with their own streets? They lace the body of California's biggest village like a concrete corset. Nowhere else are freeways knotted in such a fashion. The understandable euphoria about the relatively smooth flow of traffic (it lasted almost thirty years) has now been followed by a hangover. The stacking of the freeways today appears in a new and ominous light. We begin to see what town planners regard as the magic broom of individual traffic: freeways and parking areas eat up the last remnants of the city. "L.A. is a great big freeway," mocks the refrain of a popular song. Encaged in their cars, the freeway users are comforted by soothing words from the heavens. Pretty girls, hovering like rush-hour angels in helicopters above the traffic jams, advise the motorists on the quickest route to their office or home on behalf of a local radio station. They recommend patience: "Darlings on the Hollywood freeway! Remember that excitement is bad for your heart. . . ." In spells of acute smog their concern for the internal organs of their darlings down below may admittedly take on a desperate note: "Stop breathing. It's much better for your lungs." The commuting darlings meanwhile converse with one another by car stickers. The bumper aphorisms of the commuters speak volumes: "Remember when sex was dirty but air was clean?" "Boycott products of New York." "Drive carefully, they're waiting for your heart." "America, love it or leave it." "America, change it or lose it." "Honk if you're horny." "I'm a virgin." "Jesus saves." "Love thy neighbor, but don't get caught."

2.4 Factory bays in the dock area of Nagasaki on the Japanese island of Kyushu. A manufacturer of marine paints applies his products to the sawtooth roofs near the water's edge to test their resistance to light, weathering and the corrosive effects of salty air.

2.5 An air-conditioning installation on a high-rise building in Los Angeles. Central air conditioners on the roof supply the whole building with air through a network of distribution ducts.

2.6 The Klondike near Dawson City in the Canadian Yukon. The discovery of alluvial gold in the Klondike sparked off the gold rush of 1896. Whatever escaped the gold washers and nugget hunters of the old days is now extracted from the diminishing gold deposits by mining companies using huge floating dredges. The systematically churned waste shows where the dredges have passed.

2.7 Bingham Canyon Mine in Utah, U.S.A. This is the oldest copper mine in the world in which the ore is extracted by open mining, and the largest single mine anywhere. More copper has been extracted from it than from any other mine. It is also the biggest of man-made excavations (the excavated material of the Panama Canal would only fill one fifth of it). In short: it is the most spectacular ore mine in the world. It is 2.4 miles long and 2,600 feet deep. In winter snow falls on its rim when it is raining at the bottom. Ore trains and trucks removing the surrounding strata commute on the "benches" of this amphitheater, which are up to 50 feet high and 120 feet wide. The daily transport performance of the trains and lorries can exceed 455,000 tons of ore-bearing rock – another world record. Since 1904 three billion tons of rock have been removed from the mine and ten million tons of crude copper produced from the ore. The gigantic crater and the gargantuan loading and transport equipment are a result of the very low copper content of the ore and the unfavorable ratio of ore to country rock. Only the extraction of huge amounts of ore and strictest rationalization in its handling compensate for these handicaps. Bingham Canyon Mine belongs to the Kennecott Copper Corporation.

2.8 Breakwater in the Bay of Ise in Honshu, Japan. Because of the length of the coastline of the Japanese islands – approximately 6,000 miles – breakwaters are, not unexpectedly, in great demand. The blocks used for the stabilization of the shores come in all shapes and sizes, often in the form of composite cubes. Shore protection is not even their only purpose. In many places concrete gardens of this kind are heaped up to create suitable nesting and spawning grounds for fish.

2.9 Land reclamation near Sha Tin, Crown Colony of Hong Kong. Dump trucks are building a tongue of land into a bay of the South Chinese Sea. On it apartment buildings and a horseracing track will be erected.

2.10 Land reclamation in South Africa: wind-break fences for the tailings of a gold mine near Johannesburg. In eighty years the South African gold industry has hauled five billion tons of rock – thirty-nine times the excavated material of the Suez Canal – from depths of hundreds and even thousands of feet and crushed it to powder for the extraction of gold. The gold industry is obliged by law to protect the tailings (slime and sand dumps) against erosion by wind and water. A special organization, the Vegetation Unit of the South African Chamber of Mines, is responsible for covering these tailings with green plants. This is no easy task, as the mud and sand are extremely fine and highly acidic and contain no plant nutrients. A further difficulty is the steep slope of the spoil dumps. Before they can be covered with vegetation, they must be stabilized by a trellis of windbreaking fences made of reed. Stabilization and cultivation are done by hand, but the effort is well worthwhile: the dust clouds which previously made life miserable in the lee of the tips will soon be a thing of the past – and the green hills themselves will be available as sites for apartments, schools, sports grounds, open-air cinemas and airports.

2.11 Sculpture of the fairground: a roller coaster near Denver, Colorado, U.S.A.

2.12 Volleyball and basketball players at a high school near Santa Barbara, California. One ball is just dropping into the basket.

2.13 Competitors in a cross-country ski marathon in the Engadine, Switzerland. The road between Sils Maria and Sils Baselgia had to be prepared for the passing of the racers; by March it is in part clear of snow. The reproach directed at several of these European "people's races" ("lots of people, little race") certainly does not apply to the Engadine event. Lots of people, yes – 9,504 competitors in 1975 – but also plenty of racing: the marathon test over 26 miles long and at 5,900 feet above sea is a strenuous affair. It is not for nothing that the organizers require a medical certificate or equivalent from every competitor. The age of the participants is not limited upward, and a 77-year-old has already taken part. The record so far: 1 hour, 42 minutes, 44.1 seconds (August Broger, 1975). But in this event even the last arrival is a winner – in the private test of personal stamina and willpower.

2.14 International rowers' regatta on the Rotsee, near Lucerne, Switzerland. Final of the eights.

2.15 Mothballs for aircraft: B-52 bombers, C-124 transporters and RF-84 reconnaissance fighters (righthand edge of photograph, middle) at the U.S. Air Force base of Davis-Monthan near Tucson, Arizona. Battle-scarred or outmoded planes and spacecraft that have been eliminated from the flying inventory of the American armed forces and Coast Guard are taken over by the Military Aircraft Storage and Disposition Center (MASDC). Some of these planes are reactivated. Civil authorities may also occasionally look for something suitable among the six thousand planes of seventy different types. The forestry offices of several states use old military planes for fighting forest fires. To put these planes in the air, the MASDC takes spare parts from other craft of the same model. Empty shells and stock that cannot be disposed of are finally sold as scrap. The same things that attract retired people to Arizona are good for the planes: the climate, plenty of sun, low humidity. Another advantage is the acidfree and non-corroding soil.

2.16 Mothballs for warships: units of the destroyer class of the U.S. Reserve Fleet in San Diego, California. Box-shaped superstructures forward and astern protect the gun carriages from the sea air. Before World War II, the Reserve Fleet was satisfied if ships taken out of service did not sink. Now it has an arsenal of plastic sprays, protective coatings, drying systems and moisture-absorbing materials at its disposal in the battle against decay, mildew and every form of corrosion. The mothball fleet is moored in seven anchoring berths along the East and West Coasts and in the Pacific. It comprises some 700 battleships, aircraft carriers, cruisers, destroyers and auxiliary vessels of all types.

2.17-2.18 "The noble juice" (Shah of Persia): production plants of the oil industry. A forest of oil-well towers in the Lake of Maracaibo, below which Venezuela's richest hydrocarbon deposits lie; and billowing flames on the oilfield of Zelten, the first large field to be found in Libya. The construction of a gas liquefying plant near the oil port of Marsa el-Brega on the Greater Syrtis has now stopped this blazing waste, and the gas separated from the crude oil does not have to be flared off.

2.19 Rafts of floating timber in British Columbia, Canada. In times of critical ecological thinking the eye cannot be the sole judge of such a picture. Admittedly, almost a third of the earth's surface is clothed with forests, and as yet hardly a third of this third is being exploited. But are we to be lulled by statistics? The destructive lumbering which went hand-in-hand with the conquest of the West in the U.S.A. last century is now undisputed. Only the intervention of the federal government around the turn of the century saved the forests of the Rocky Mountains and the Pacific Coast Ranges. Luckily, this incredible shortsightedness in the past has sharpened the consciousness of North Americans for the importance of the forests. It has thus helped spread awareness of the fact that a timber trade that does not equate felling rate and growth rate is sawing off the branch on which it sits. But what of other parts of the world – the immense wooded regions of Asia, Africa and Latin America? Here complete clearing of the tropical rain forests heralds an ecological disaster.

2.20 A supertanker under construction in Nagasaki, Japan. The "Onyx" lies in the slipway, awaiting launching in a few weeks. With a length of 1,050 feet, a beam of 177 feet and a height of 85 feet, she has a carrying capacity of 268,951 tons and belongs, in the jargon of tanker shipping, to the category of the Very Large Crude Carriers (VLCC). She is an "oilephant" but not a "mammoth of the sea" (Ultra Large Crude Carriers of over 350,000 tons deadweight). The gantry crane bears the emblem of the construction firm Mitsubishi; owner of the ship is the French Compagnie Navale des Pétroles. Fifty percent of the world's ocean-going tonnage is accounted for by tankers; in mid-1975 there were more than 500 VLCC and ULCC ships. The only limit to jumboism in shipbuilding has long been set by the size of ports, straits and other shipping channels. The latest advances in automation and remote control, loading and unloading equipment, navigational and supervisory aids, seemed to preclude the possibility of accidents resulting in oil losses. The impossible nevertheless promptly occurred: fires, explosions and huge patches of oil on the sea made a mockery of the planners' claims. Supertanker presumption has thus run aground. Even for a standardized, less euphoric development of supertanker shipping there are now, after economic setbacks, obstacles not shown on any nautical map. But the braking distances in this branch of industry are long. A VLCC giant traveling at a speed of only six knots (just under seven miles per hour) takes just under one mile to stop with brakes fully applied.

2.21 Main station, Zurich, Switzerland: marshaling yards for passenger coaches. The bridges supporting the overhead contact wires divide up the picture. In a 24-hour day, 3,500 passenger coaches are made up into 600 trains in the passenger station of Zurich.

2.22 Color-coded containers parked in the container port of Hong Kong. Containerization has revolutionized the handling of goods in less than a decade. The goods are stowed in loading units that can be placed on rolling stock and lifted by cranes. The containers do not go from port to port only, but from loading station to destination. Together with specialized transport, loading and unloading equipment they have had a rationalizing effect which also pleases the eye. At present the containers are available in two standard sizes, the smaller unit being twenty, the larger forty feet long. More than a million of these units are currently traveling on oceans, rails and roads. An optimistic estimate is that world transport will need half as many again within the next five years.

2.23 A kind of large-scale Paul Klee: bridge building site near Santa Barbara, California. Old carpets, which are thoroughly soaked at least once a day, keep the freshly concreted fairway moist. When concrete hardens, gravel and sand are bound to the cement, which undergoes a gradual chemical reaction. At this stage, water is needed and heat is given off. In order to obtain good concrete, the builder protects it during this phase from loss of water and rapid cooling. Direct solar radiation and drafts are particularly harmful and may cause shrinkage cracks. The system adopted in California, though picturesque, is today out of date. In Europe special mats – a Swiss development – have come into use. They retain the heat developed as the cement hardens and make it unnecessary to keep wetting the concrete. This new procedure is more economical but optically rather dull – the so-called Guritherm mats are at present only available in green.

3.1 The two craters in the caldera of the Erta Ale volcano in the Afar depression. The diameter of the bigger crater is 980 feet, that of the smaller one 230 feet. Erta Ale is, technically speaking, a basaltic shield volcano in the process of formation. The picture was taken in 1965, when the lava lake was at low tide. In the bigger crater the level of the lava is approximately 460 feet below the crater rim; steam and smoke prove that there is also a lava lake in the smaller crater.

3.2 Erta Ale in spring, 1974: the big crater has filled and narrowed into a magma lake 165 feet in diameter. The molton lava forms a thin skin where it is exposed to the air. Where this skin breaks, hell blazes and bubbles.

3.3 Cone of an inactive volcano in the Erta Ale range. It was formed when the Afar was still an arm of the Red Sea; by the time it was left high and dry, it had already become dormant. Fish live in the crater lake, and the fringe of vegetation indicates that not even the blistering summer sun can dry it up completely. A natural cistern collecting the rainfall? Hardly, since rain here falls scantily and sporadically. There must be some connection with groundwater reserves; only an underground water supply can make up the losses due to evaporation over long periods of time. Prospectors baptized the structure "The Colosseum." Later it received the more official name of "Mount Asmara."

3.4 A curious ring consisting of multicolored rock salt and fragments of basalt. The geologists take it for a salt plug encased in rock, which surged up from the depths and broke through the surface. Its present structure is the result of erosion by rainfall and floods. The Afar people attribute special healing powers to the salt of this plug.

3.5 A stillborn volcano, so to speak: a magma intrusion from the depths of the earth which did not reach the surface but threw up a whirling labyrinth of salt blocks in the overlying strata. The striking rectangle at the center could be the remainder of a sealing block. Erosion transformed the labyrinth to its present state by washing away the more easily soluble salts and thus filling the spaces between the blocks up to the level of the surrounding, hexagonally textured salt plain. The "vortex" has a diameter of 490 to 655 feet. My attempt at interpretation here is based on information obtained from the geologist Derek T. Harris, who was a member of the geological team that investigated the potash deposits of the Afar depression, estimated at hundreds of millions of tons. Harris had never actually seen the vortical labyrinth (which was only six miles away from the prospectors' base on the hill of Dalol) but he associated my aerial picture with several enormous magnetic anomalies of similar shape appearing on the magnetic map of the area. These anomalies point to intrusion bodies with a high iron content, many of which, however, have produced no sign on the desert surface. Harris planned a hoax with my picture: he wanted to publish the photograph of the labyrinth in a reputable scientific journal as an archaeological site, as vestiges left by Man. His premature death prevented him from carrying out the plan. (His hoax might easily have backfired: James G. Holwerda, another geologist with years of Danakil experience and a colleague of Harris, seriously believes this vortex to be the remains of a camp of salt gathering Danakil nomads. Holwerda, too, judged from the aerial picture. Only an actual inspection of the site can clarify the matter.)

3.6 A bizarre world of pure salt: the hill of Dalol in the northern part of the Afar. It rises 150 feet above the salt plain lying nearly 400 feet below sea level. Magma surging from underground thrust up the hill, which is dotted with geysers; occasional rainfall has melted the salt blocks to towers and needles. Caps consisting of gypsum, which is harder to dissolve, protect the tops of the salt excrescences.

3.7 A tumult of colors in the salt desert: the rounded summit of the hill of Dalol. Hot springs (at about 212°F) shed their salts; the colors – including yellow, which might suggest sulfur efflorescence – are due to the oxidation of small iron admixtures. Each spring flows for approximately one month; once it has dried up, the deposits gradually fade.

3.8 3.9 Springs at the edge of the Afar depression, where the groundwater horizon cuts the slope. The groundwater is fed by the rainfall in the Ethiopian highlands as well as by underground water emerging along fault lines. The latter can be recognized by a temperature above the ambient level.

3.10 The monsoon rains in the Ethiopian highlands and the very scanty local precipitation bring some water to the Afar depression. The traces of the run-off water are engraved in the black "desert varnish," a crust of manganese. The diagonal running across the picture, which at first looks like a scratch, is in reality a caravan trail – trace of Man among the veining of natural forces.

3.11 A salt caravan on its way back to the plateau. Exploitation of the salt deposits by hand has been for centuries the only source of income for the Danakil tribes in the north of the Afar triangle. The lowest point of the salt pan never dries out completely, and the monsoon rains over the highlands nourish it anew each year. Twenty thousand pack animals – camels, mules and donkeys – journey to and fro on the ancient salt road. The annual output is about 17,000 tons of table salt.

4.1 Two center-pivot irrigation systems between North Platte and Imperial in Nebraska, one of America's most important farming states. In the dark halves of the circles alfalfa is growing, in the light ones corn. The farmer can select any combination of crops; his choice is limited only by the nature of the soil. Different moisture requirements of crops in the same circle are no problem with a center-pivot irrigation system. The swivel arm can be programed, for instance, to turn back after half a revolution.

4.2-4.3 Fields lying fallow near North Platte. In order to prevent the growth of weeds, improve the moisture content of the soil and counteract wind erosion, the uncultivated field is repeatedly worked during the vegetation period. The regular texture in the light areas has been left by a sweep. The farmer is now working with a tandem disk harrow, always driving across the slope of the terrain. The decorative border of the stubble field is produced by the turning sweep. Wheat was harvested here last season.

4.4 The farm of William Sturtevant, near Wauneta. It comprises 259 hectares (640 acres) of Sturtevant's own land, plus half as much again that he has on lease. Terrace cultivation is practiced to save water. The crops are rotated: wheat, corn and fallow, in that order. This photograph was taken in September. Wheat had already been sown on the summer fallow (brownish shades): on the dull green fields corn is ripening; and the harvesting of sorghum – brighter green and newly reaped fields – is in progress. To the north of the farm (left margin of the photograph) the Badlands begin; their runoff flows into a creek called Stinking Water, which flows into Frenchman River, a tributary of Republican River. Where the Badlands have a grass cover, they are used for pasture. Sorghum and some oats supply the winter fodder for the cattle, which also graze the corn stubble.

4.5-4.8 Works in an art exhibition – or "only" field graphics? For every line the farmer draws, for every pattern he creates, he has a good reason. Sometimes, admittedly, it is only an excuse. While the farmer was working with a disk harrow on a wheatfield lying fallow for the summer, he was forced by rain to interrupt his work for several days (Plate 4.7).

4.9 Growing wheat along contour lines, known as "contour farming." The dark fields are under wheat; between them are fields of wheat stubble which are lying fallow for the summer. They show traces of having been worked with large sweeps; which separate the roots from the stubble but leave the latter standing to protect the fields from wind erosion.

4.10 The agricultural landscape near Scottsbluff, early June: strip and terrace cropping of wheat and alfalfa. The light strips are lying fallow or planted with sugar beet which has just sprouted. The strips lie at right angles to the direction of the prevailing north-west wind. The direction and size of the strips are also influenced, however, by the terrain, the condition of the soil and the farmer's mechanical equipment. The two irrigation circles, where corn is being grown, are new and the swivel arms and sprinkler nozzles have not yet been mounted. Each square enclosed by the roads is a "section," one square mile in area.

4.11 A farm near Imperial. It consists of a total of fifteen center-pivot irrigation systems, each of which occupies one-quarter of a section. (The same farmer owns two further farms of a similar size.) Corn is growing on the light circles, grass and alfalfa on the dark ones. On two the

alfalfa is just being harvested, while in a further circle – a segment of which is visible at the bottom of the picture – the farmer is growing potatoes.

A center-pivot irrigation system growing alfalfa. Exactly 4.12 controlled irrigation permits up to four harvests per year. At full speed (one revolution in twenty hours) and with a water supply of 5,000 gallons per minute, a standard ten-limb system supplies about 0.17 inches of water to an area of 135 acres. At very slow speed (one revolution in two hundred hours), the amount of water can be increased almost tenfold. The systems run well above ground and can be used even with tall plants – in fact, corn is grown under every second irrigator in Nebraska. The supports of the swivel arm describe concentric circles in the field. The well and pumping installations are outside the photograph, a short distance from the center of the system. This new irrigation technique has also been a success in other countries. Among the users are Mexico, Australia and Libya; the last-named is using the technique to exploit extensive groundwater supplies near the oases of Kufra, in the midst of the desert.

A bare fallow field – previously planted with corn or sugar 4.13 beet – has been worked over with a rotary hoe to prevent the blowing away of the topsoil. Evidently this measure has been only partially successful, for the aerial picture shows traces of drifting soil here and there. The farmer hopes for rain. If he is disappointed, he will have to have recourse to other agricultural machines from his defense armory to deepen the furrows or raise their edges.

5.1 The Jebel Musa ("Mountain of Moses"), a summit in the granite wilderness of South Sinai. Rising 7,500 feet above the level of the Red Sea, it is not the highest mountain of the area, but for sixteen centuries the Christian tradition has venerated it as the holy mountain upon which Jehovah appeared to Moses and made his covenant with the tribes of Israel. Snow fell on the day before the photograph was taken – a sprinkling only, but unusual even for a December day.

5.2 The ruin-strewn hill of ancient Jericho (Tell es-Sultan), West Jordan. The whole of the hill, which is sixty-five feet high, consists of the rubble of earlier civilizations. No other town on earth can compete with Jericho in age. In the deepest layer, opened up by the exploratory trenches visible on the photograph, were discovered the oldest stone houses and town fortifications known so far – dating back to the eighth millennium BC. Fundamentalists who claim that the Bible cannot be wrong are admittedly out of

luck in the case of Jericho. Around the time of the occupation of Canaan by the Israelites (between the fifteenth and thirteenth century BC) Jericho, though already over six thousand years old, would seem from the excavations to have declined to insignificance. In the previous seven hundred years, Jericho's walls are known to have been destroyed and rebuilt – or at least mended – no fewer than twenty times. But of the walls which, according to the Bible, Joshua's trumpets brought tumbling down the archaeologists can find no trace.

Et-Tell near Deir Diwan, in the mountains of Judaea, West 5.3 Jordan. The ruins are generally thought to be the remains of Ai which, according to the Biblical report, was the first town Joshua took in the mountain country, when he "made it an heap for ever." The excavation evidence contradicts this version. When God's chosen people settled in Canaan, the town had been lying in ruins for a thousand years. The contradiction might, of course, be due to the fact that the identification of et-Tell with Ai is erroneous.

5.4 The hill with the ruins of the Philistine town of Ekron, Israel. Spring flowers effectively set off the tell against the surrounding cultivated land; the rectangular walls of the acropolis shine through the flower carpet. According to the Book of Judges, Ekron fell after the death of Joshua; but it was conquered, if conquered at all, only temporarily. For a short time the Philistines even put up in Ekron the Ark of the Covenant, which they had captured. Later, after Goliath's death, they sought refuge in the safety of the town. Ekron only came permanently into the hands of the Israelites in the second century BC, when the Syrian king Alexander Balas presented it as a gift to Jonathan Maccabaeus. Modern excavations have yet to be undertaken in Ekron.

5.5 A town without its like: Jerusalem. Not the heavenly Jerusalem the soul seeks for, but the Jerusalem of history conquered, torn down and rebuilt, a bone of contention to this day, earthly and in the news. The rays of the setting sun over the old part of the town light up, on roofs already in the shade, the forest of television aerials which sprang up after the war of June 1967 and the reunion of the divided town. King David took the Jebusite fortress on the heights of the Judaean hills and from there ruled over north and south; Solomon made the town rich, Herod made it great. For nineteen centuries after the destruction of the second temple it remained a focal point of hope for dispersed Jewry: "Next year in Jerusalem!" The Christian churches, too, see more in Jerusalem than just a hotly contested town at a crossroads of history: here, where God's son died and rose from the dead, time and eternity meet and merge for the believer. And even this is not the end of its holiness: Mohammed, when in flight, sometimes told his disciples to turn their faces to Jerusalem instead of Mecca in prayer – and it was from the temple hill that the prophet rode up to heaven to receive illumination from Allah. The present-day walls were built by Sultan Suleiman the Magnificent. In the southeast (top right in the photograph) can be seen the Haram ash-sharif, the former temple square, where since the seventh century the Dome of the Rock stands (over the rock of Abraham's sacrifice and of Mohammed's heavenly vision) and since the eighth century the Aqsa Mosque. The evening sun falls on the Wailing Wall, a part of the western retaining wall of the Herodian temple, which has been freed of its debris since 1967 and is now accessible across a wide square. At bottom right in the picture is the complex of Christian sanctuaries, above all the place of Christ's crucifixion and burial (under the domes of the Church of the Holy Sepulchre). Of the ten measures of beauty that descended on the earth, Jerusalem claimed nine. But the Babylonian Talmud says that nine is also Jerusalem's share of the ten measures of sorrow. And politicians have done everything in their power to preserve the rightful proportions – if not those of spiritual beauty, then at least the town's share of its only too palpable sorrow.

The tell of the town of Megiddo at the western entrance to 5.6 the plain of Jezreel, Israel. Excavations have exposed palace ruins, fortifications, a water supply system designed for times of war, stables and storehouses. Solomon fortified Megiddo – made it a district capital and garrison for chariot troops – so as to control the caravan routes between the valley of the Nile and Mesopotamia.

The acropolis of the town of Hazor, Israel. The road from 5.7 Tiberias to Metulla runs round the venerable tell on the eastern edge of the upper Galilean mountains. Hazor was the capital of the northern Canaanite territories; the Israelites conquered the town, and Solomon turned it into a fort. In the upper half of the picture are citadels, ruins of royal storehouses and, between them, the rectangular shaft that gave access to the drinking water.

The ruins of the town of Gezer, on the edge of the hill 5.8 country in Israel between Jerusalem and the sea. A sap – reminder of recent hostilities – marks the northern and eastern edge of the hill. Above the center of the picture, on the left, the ruins of a monumental gate from the time of Solomon; and to the right of it, where archaeologists have filled the sap with excavation debris, a casemate of the inner fortifications of the town and a square with *mazzeboth* (stone pillars) which may commemorate a solemn alliance. Gezer, a royal residence of the Canaanites from the third millennium BC, controlled the routes between Egypt and Assyria. Solomon obtained it as part of the dowry of the daughter of Pharaoh whom he took into his harem. He fortified the town against Ekron (Plate 5.4). The famous calendar of Gezer is a limestone tablet found there, presumably dating back to Solomon's times and listing the twelve months with information about sowing and harvesting.

The escarpment of the mountains of Judaea looking to- 5.9 ward the Jordan depression and the Dead Sea: the wilderness of Judah with the old road to Jericho. The gorges and caves of this region have offered shelter to the banished and persecuted, to preachers and robbers, pioneers and footpads throughout Palestine's tumultuous history. At the top left edge of the picture, behind the haze, the oasis of Jericho and the Dead Sea.

Samaria on its mountain, West Jordan. Samaria is one of 5.10 the very few towns founded by the ancient Israelites. Omri, one of the most enterprising kings of the northern territories (though probably of Arab extraction), built Samaria as his capital around 870 BC. Excavations have revealed ruins of the Israelite king's palace on the mountain and town fortifications belonging to the same epoch. The forum, a basilica, a theater and a columned street – from which the archaeologists have also removed the debris of the millennia – recall the town's golden years under Herod and the Romans. Outside the left-hand edge of the picture lies the Arab village of Sebastiye. Its name perpetuates a piece of flattery by Herod, as under his rule Samaria was rechristened Sebaste from Sebastos, Greek for Augustus, Herod's imperial benefactor.

5.11 The hill country of Samaria, West Jordan. In the Old Testament this central region of the West Jordan highlands was known as the mount of Ephraim.

5.12 The Herodeion on the threshold of the wilderness of Judah, West Jordan. Herod maintained a palace and baths on the mountain during his lifetime, and it is probable that he intended the place to be his monument from the first. Between 22 and 15 BC he had the natural hill remodeled and banked up to form a regular truncated cone, probably wishing to imitate the mausoleum built a few years previously for the Emperor Augustus. In the Jewish war of liberation against the Romans the Herodian fortress was one of those that still held out against the besiegers when Titus and Vespasian, after taking Jerusalem and destroying the temple, were celebrating their triumph in Rome.

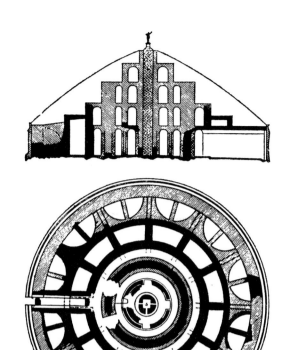

The mausoleum of Augustus (after *Segal*).

6.1 Ayers Rock in Central Australia. The rock haunts popular tourist literature as "the biggest monolith on earth." Not only is it not the biggest, but it is not a monolith at all: rather an inselberg of stratified rock (arkose). For local tribes of the Australian aborigines, Ayers Rock is a mount of the gods, totally enveloped in legends and myths. Caves, shelters and other minor features of the rock are associated with memories of cultural heroes of the "Dreamtime," and sacred rites continually make this time operative in the present. The first and last rays of the sun set the mountain on fire – one can hardly put it otherwise. Its spectacular color changes have made this rock in the red, dead heart of the continent into more than just an ordinary destination for sight-seers, even for white Australians, and an excursion there takes on the dimension of a pilgrimage.

6.2 The Church of St. George Beta Giyorgis in Lalibela, Ethiopia. Lalibela's monolithic churches rank high in the catalogue of the wonders of the world. Cut from a single rock, yet reproducing architectural forms in their internal and external design, they are in fact huge sculptures – architecture by subtraction. They stand at the bottom of

pits, trenches having been dug around them to allow work to be done on the living rock. It is assumed that Beta Giyorgis, the last of Lalibela's monoliths, was built at the beginning of the thirteenth century. Its elegance is unmatched. The cruciform building, with three crosses inside each other as a roof decoration, stands on a three-stepped cross-shaped base. The church above this base was carved from a rock thirty-five feet high and forty-one feet square. Passages lead down to the bottom of the excavation. The church interior is a cruciform space undivided by pillars.

Mont-Saint-Michel in the Département of Manche, 6.3 France. Low tide has exposed the sandbanks around the granite island. This innermost niche of the Gulf of Saint-Malo has continental Europe's largest tidal amplitude (up to forty-six feet); at neap tide the sea uncovers up to nine miles of sandy flats. Despite its uneven course, the tidal inlet at the foot of the hill was for centuries the boundary between Brittany and Normandy. Water engineering measures at the beginning of this century, however, robbed the Bretons of "Le Mont," to their persisting mortification. The presence of monks' settlements on the granite

123

crest can be traced back as far as the eighth century, and its Benedictine history begins in the tenth. The buildings with the monastery church that crowns them were erected between the eleventh and sixteenth centuries. No besieger has ever conquered the abbey. Nevertheless, this prayer in architecture has known times of humiliation. Temporarily it served as a prison; later it became a museum. Since 1966, three monks have been living on Mont-Saint-Michel again. They assemble speedometers for a nearby factory.

6.4 The sacred stone circle of Stonehenge in Wiltshire, England. An African statesman not long ago wished (rhetorically) to take Stonehenge home with him as a proof that even Europe had gone through primitive stages of culture. However undisputed this statement is, Stonehenge is far from being a good illustration of it. Its concentric circles with their two inscribed horseshoes were created by what was then – the first half of the second millennium BC – a monumental effort. Its opening toward the point of sunrise at the time of the summer solstice, together with the alignment of the stones according to certain solar or lunar phenomena, showed Stonehenge to be a center of sun worship. In the 1960's, a Boston professor named Gerald S. Hawkins offered a new solution to this petrified conundrum from European prehistory. A modern high-speed computer helped the astronomer in his surprising new assessment. The stone monument, said Hawkins, is itself a sort of New Stone Age computer, with the aid of which astronomer-priests were able to predict exactly the succession of the seasons and the eclipses of moon and sun.

6.5 The spiral minaret of the Great Mosque of Samarra, Iraq. The ramp spirals up in five coils from the base to a cylindrical top story. The base measures 108 feet square, and the topmost platform is 164 feet above the base. During the siesta natives come here for cooling shade. Caliph al-Muta-wakkil (847–861), who was unrivaled as a builder in Samarra, took up the Sumerian-Assyrian-Babylonian tradition of the sacred tower. The spiral minaret is accordingly an early Islamic echo of the Tower of Babel. With it the Caliph boldly countered the misunderstanding caused by the biblical polemic and by a similar allusion to the Babylonian tower in the Koran. The ziggurats were never fists rebelliously shaken against heaven; instead, the temple towers were an invitation to God to descend and live among needy mortals. The circular shape of the Malwiya ("the spiral") stresses the desire for cosmic unity.

6.6 The Tower of Babylon, Iraq, on the site of the ruined city. This picture shows the results of plundering of the coveted fired bricks by local inhabitants in the nineteenth century. Groundwater instead of burnt bricks now encloses the central part of the ziggurat, which is made of unburnt clay bricks. The tower is today disrespectfully called "the saucepan"; the handle is formed by the trench of the old stairway. The date of the founding of the ziggurat of Babylon – called Etemenanki ("foundation stone of heaven and earth") – is unknown, as its lowest parts lie in the groundwater and cannot be excavated. The oldest archaeological finds date from the reign of Esarhaddon (680–669 BC); the florescence of the sanctuary began with Nebuchadnezzar II (604–562 BC), the sacker of Jerusalem. In this New Babylonian version of the ziggurat the Sumerian and Assyrian patterns for a temple tower are combined. From the two stories reached by a huge outdoor staircase, a spiral ramp led up to four further floors. The ziggurat of Babylon was the only tower in Mesopotamia in which the height was equal to the length of the sides (300 feet). Xerxes reduced the tower to ruins, and Alexander the Great began to demolish it in order to make room for a new structure – a project that was cut short by his death.

The inner shrine of Ise on the Japanese main island of 6.7 Honshu. The temple precincts of Ise are the most revered of all the sanctuaries of Shintoism. Only a century ago Buddhist nuns and priests were not allowed to enter it, and it is still without the Buddhist statues, incarnations of Shinto gods, that are found everywhere else. The holy of holies is trebly guarded by fence and palisade; it comprises treasure chambers for cultic objects and liturgical robes and the main hall in which lives the sun goddess Amaterasu, ancestress of the imperial house, who is symbolically represented by the sun mirror, holiest of the imperial insignia. The gateway to the inner shrine is opened only to high priests and the imperial family and its messengers. Tradition, however, has nothing against inspection from above. The Amaterasu shrine dates back to the beginning of the Christian era; since the seventh century it has been completely renewed, including the secondary buildings, every twenty years, if the times permitted. The sixtieth renewal took place in 1973, an exact copy being made of the old building and thus of all the previous buildings. Wood and straw are used exclusively, without either nails or mortar. The old building is divided into its component parts and sold down to the last chip on the devotional market; its wood is even reduced to toothpicks. In this way it pays for the new building.

A stupa in Sri Lanka (Ceylon). The stupa (the Singhalese 6.8 usually call it a "dagoba"), which is now the essential form of Buddhist religious architecture, developed from the pre-Buddhist princely tomb of ancient India. The Enlightened One himself instructed his disciples to honor kings and wise men with stupas. After his death, eight stupas were erected over holy relics of the Buddha, but King Asoka alone (third century BC), a pious champion of Buddhism, multiplied their number to tens of thousands. Built mostly over a holy relic, the stupa is always architectural sculpture without an interior. Above the stepped base, placed in an enclosure with four gateways, rises a hemisphere which may also be bell-shaped or bulbous; the cubic superstructure at the top (sometimes containing the reliquary) is crowned by a cylinder and cone. This feature represents an umbrella – an ancient Oriental symbol of a ruler. To the believer the stupa symbolizes Parinirvana, the final, perfect Nirvana, the state of absolute salvation.

6.9 The radar radiotelescope of Arecibo, Puerto Rico. Its reflector, the largest on earth with a diameter of some 985 feet, is a stationary installation set in a depression in the terrain and directed toward the zenith. In contrast, the antennae on the triangular platform 525 feet above the ground of the punchbowl are swivel-mounted. Thus the apparatus can "see" 40 percent of the visible vault of the skies. (At the time this photograph was taken, the reflector consisted of fine-meshed wire netting; since then it has received an aluminum coating to improve its geometric definition). When Columbus set out to prove that the earth was round, he discovered Puerto Rico. Today astronomers penetrate to the limits of the universe from the tropical Antilles. The radio observatory of Arecibo records the murmurs of the quasars and pulsars. Nearer home, *El Radar* explores the ionosphere and the planets, whose orbits, thanks to Arecibo's tropical situation, are almost always in the field of vision of the telescope. In Arecibo it was shown that Venus, unlike the other planets, turns clockwise and not anti-clockwise. And that Mercury rotates on its axis in a mean period of fifty-five days – and not as previously believed, in eighty-eight days. Carl Sagan also uses the instrument to search for radio messages from extra-terrestrial intelligences.

6.10 The Teatro Amazonas in Manáus, Brazil. Manáus on the Rio Negro, whose purple waters mingle with the brown floods of the Amazon nine miles downstream from here, did all in its power around the turn of the century to forget that it was surrounded by primeval forest. London provided the gentlemen's tailors, Paris the jewelers; cocottes and carriages were Continental. One of the trading kings in the jungle town even had his laundry sent to Europe by ship to be washed and ironed. Rubber was a monopoly of the Amazon, its prices were dictated by the dealers in Manáus – and world demand was rising. The spectacular theater, a copy of the Opéra in Paris, was the crowning symbol of the *dolce vita*. Caruso sang in it, Sarah Bernhardt acted there, and Pavlova danced. The big money also attracted lesser talents 800 miles up the Amazon, at that time an adventurous, fever-menaced journey. But even before World War I the rubber plantations of Southeast Asia had dethroned the collected rubber of the Amazon. Manáus went on for a while as if nothing had happened – but before long the tropical air with its smell of decay took

over again where the fragrance of French perfumes had once prevailed.

6.11 The collegiate church of the Benedictine abbey of Maria-Einsiedeln, Switzerland, one of Europe's finest Baroque monastery complexes. The first monastery was built here by the Benedictines in the tenth century. The monastery and the pilgrim-frequented church were built in their present form in the eighteenth century – the sixth successive structures on this site. The fountain of St. Mary in the middle of the square is answered by a statue of the Madonna on the front gable.

6.12 Earth Art cut into the American earth: "Double Negative" by Michael Heizer on the edge of Mormon Mesa in Nevada. The cut is 30 feet wide, 50 feet deep and 1,640 feet long – including the middle part, which belongs to the concept. The tongue-shaped rubble dumps are also an integral part of the work. "The greatest sculpture in the history of Western art" – and its turning point too – a prophet of Earth Art predicted in a reputable art journal. Nothing less than a record will do – and only fifty-five miles from Las Vegas; the West is in any case a courtesan who is ready to sleep with every art historian. "Double Negative" is undoubtedly the most labor-intensive work of Earth Art to date. Heizer, born in 1944, drilled, dynamited and bulldozed it in 1969/70 in a mountain he had bought a piece of. He used two tons of dynamite and moved over 200,000 tons of rock. The earth artist rejects the compulsions of the art trade – escapes from studios, galleries and museums and liberates his work from the pull of the consumer society. Heizer says: "One of the implications of Earth Art might be to remove completely the commodity status of a work of art and allow a return to the idea of art as . . . more of a religion."

6.13 "Spiral Jetty" by Robert Smithson, another American land artist, in Utah's Great Salt Lake. Smithson, born in 1938, heaped up his spiral jetty in 1970 on a piece of land he had leased for twenty years. He kept account of exactly how much material he tipped into the lake to make the jetty, which is 490 feet long and 15 feet wide: 6,000 tons of basalt blocks and riparian rubble, 292 truckloads in all. This mysterious sign in the desolate landscape at the northern end of the lake reacts like an organic entity to the changing seasons. The photograph shows the spiral at high water. When the lake falls as a result of summer evaporation losses, the spiral is left dry, and the black blocks become encrusted with white salt crystals. At the same time the red coloring of the water is intensified, as the precipitation of the salt leads to an explosive multiplication of red algae. It should be added that the photograph does not do complete justice to the intentions of the artist. The sign value was secondary to the experience Smithson had when walking along the spiral from the outside, of the

125

swirl and suction of life, which spirals in irreversibly and ever faster toward the point of final truth – death. In 1973, when Smithson was searching from the air above Texas for a site for a new earth composition, a spiral ramp, he crashed to his death.

6.14 Cemetery of well-to-do Chinese near Medan, Sumatra, Indonesia. The striking and studied irregularity in the direction of the graves is the work of the geomancer, a wise man skilled in *feng-shui*. In the upper right-hand corner a grave is just being dug. The bigger tombs are family vaults where wife and concubines lie beside the owner. In front of the horseshoe-shaped tumulus covering the coffin there is space for funeral rites. Tombs of this kind extend over whole mountainsides in China. They reproduce, on a reduced scale, the ancient monumental tombs of the imperial house and of prominent persons. The mound symbolizes the mountain on which the soul of the deceased finally quits the earth to soar aloft to heaven. Thus the questioning never ends; the survivors go on asking in the name of the dead.

Author's note

The Photographs

To collect the photographic material for this book, the author spent more than a thousand hours in small chartered planes flying over fifty-nine countries and territories. In the final selections of the illustrations, subject matter came first and geographical variety was only a secondary consideration. Even so, the examples included represent more than twenty-five countries and territories on six continents.

Equipment

Nikon cameras equipped with motor drive and Nikkor lenses with focal lengths from 28mm to 180mm were used. The motorization of the camera made it possible to select subjects and angles in fractions of a second while approaching and flying over the target. This resulted in a considerable saving, as expensive helicopters were not needed.

Film Material

Ilford FP-4 was used for the black-and-white photographs, Kodachrome II or Kodachrome 25 for the color shots. The resolving power and sharpness of Kodachrome films, together with modern reproduction techniques, yield results in the miniature (35mm) photograph which make cameras of the next larger negative size unnecessary.

Publication Permits

The Austrian Federal Ministry of Defense released Plates 1.5 and 1.6 for publication with the note Z1.10.831-RabtB/74.

Literature Used

The titles listed below are meant as an expression of the author's gratitude to writers whose books and papers helped him in the compilation of the notes on the photographs – they are not a bibliography, not even a select one.

Where the information given on the photographs is widely known or easily accessible, references to the literature have been omitted altogether. Persons who assisted the author by supplying information orally or by letter are listed, with thanks, in the Acknowledgments on page 128.

A ROOF OVER ONE'S HEAD
Le Corbusier: *Aircraft* (*The New Vision*, Vol. 1.) London, 1935. Hofer, P., "Die Stadgründungen des Mittelaters zwischen Genfersee und Rhein," in *Flugbild der Schweiz*, Berne, 1963. Westphal-Hellbush, S.: *Die Ma'dan*, Berlin, 1962.

CALLIGRAPHY OF THE INDUSTRIAL AGE
Cook, W. H. "Stabilizing Gold Mine Slimes, Dams and Sand Dumps by means of a Vegetative cover." Lecture to the South African National Association for Clean Air, 1970. (Available through the Chamber of Mines Services, Johannesburg). Groves, J. E.: "Reclamation of Mining Degraded Land," in *South African Journal of Science*, Vol. 70 (October, 1974).

FLIGHT OVER THE AFAR
Holwerda, J. C. and R. W. Hutchinson: "Potash-Bearing Evaporites in the Danakil Area, Ethiopia," in *Economic Geology*, Vol. 63, No. 2 (March/April, 1968).

BIBLICAL SITES AND CITIES FROM THE AIR
Dever, W. G. (*inter alios*): "Further Excavations at Gezer, 1967–1971," in *The Biblical Archaeologist*, Vol. 34. Cambridge, Mass. (1971). Segal, A.: "Herodium," in *Israel Exploration Journal* Vol. 23 (1973).

MONUMENTAL QUESTION MARKS
Hawkins, G. S.: *Stonehenge Decoded*, New York, 1965. Killer, P.: "Leitfossilien der Zeitkunst 1945–1973," in *DU*, April 1973. Tomkins, C.: "*Onward and Upward with the Arts,*" in: *The New Yorker*, February 5, 1972.

Acknowledgments

My thanks go first, and with particular emphasis, to Swissair, and above all to its advertising manager Albert Diener and its art director Fritz Girardin. Commissions for aerial photographs for advertising and promotional use were only one form of encouragement I received from them; the second form derived from a shared conviction that the view from above could contribute a great deal to Man's understanding of himself. Without Swissair's stimulation this book would have got bogged down in its early stages.

My thanks go to the photographer Emil Schulthess. As the designer of a number of posters and calendars for Swissair using my aerial photographs, he unselfishly helped to focus the picture themes more sharply. Collaboration with him made my own eye more sensitive, and the book owes a lot to the standards he set.

My thanks go to the designer Hans Frei. Not shrinking from the unrewarding role of fill-in, he stepped in at a critical stage of the printing and made his immense knowledge available for the color corrections in the plate section.

My thanks go also to the National Geographic Society in Washington DC. The Society enabled me to undertake a reconnaissance flight along the Niger and permitted me to include some of the resulting photographs in this book.

My thanks go finally to all those who assisted me in the preparation, execution and evaluation of flights, namely Gianni S. Amado, Dakar; Ali Ghalib al-Ani, Baghdad; Emil Baettig, Mexico City; Behnam Abu's-Suf, Baghdad; Rainer Michael Boehmer, Baghdad; Valerio Bondanini, Rome; Theodor Bregger, Milan; Helmut Brinker, Zurich; Peter Bürgi, New York; Donald G. Campbell, Sydney; René Caretti, Bucharest; Arnold Catalini, Baghdad; W. H. Cook, Johannesburg; Reinhold Debrunner, São Paulo; Pio G. Eggstein, Johannesburg; Otto Eichenberger, Zurich; Peter Eigenmann, Hong Kong; Michael Evenari, Jerusalem; Catherine Flynn, Gove; Goro Kuramochi, Tokyo; Guy G. Guthridge, Washington DC; Derek T. Harris, Los Angeles; Carlos Häubi, Mexico City; Michael Heizer, New York; Sebastian Hirschbichler, Vienna; Thino W. Hoffmann, Colombo; James G. Holwerda, Singapore; Max Hunziker, Zurich; Thomas Immoos, Tokyo; Zacharias B. Kaelin, Sydney; Peter Kaplony, Zurich; Max A. Landolt, Washington DC; Fritz Ledermann, New York; Rudolf Müller, Tokyo; Walter Müller, Lagos; Peter R. Odens, El Centro, California; Erwin Walter Palm, Heidelberg; Roger Pasquier, Peking; Fritz Peyer, Hong Kong; Hans Röthlisberger, Uerikon/Zurich; Hans Rüesch, Medan; Alfredo Schiesser, Philadelphia; Walter A. Schmid, Zurich; Hans-Jürgen Schmidt, Baghdad; Ernst Schumacher, Berne; Mario Selva, Athens; Adolf Senn, Rome; Robert Smithson, New York; C. G. G. J. van Steenis, Leyden; William Sturtevant, Wauneta, Nebraska; Elva Weakley, Calexico, California; August E. Weber, Hong Kong; Max Zimmerli, Kinshasa; Pierre Zoelly, Uerikon/Zurich.

128